T0361803

Partial Differential Equations for Mathematical Physicists

Partial Differential Equations for Mathematical Physicists

Bijan Kumar Bagchi

CRC Press
Taylor & Francis Group
Boca Raton London New York

CRC Press is an imprint of the
Taylor & Francis Group, an **informa** business

CRC Press
Taylor & Francis Group
6000 Broken Sound Parkway NW, Suite 300
Boca Raton, FL 33487-2742

International Standard Book Number-13: 978-0-367-22702-9 (Hardback)

Library of Congress Control Number:2019946208

**Visit the Taylor & Francis Web site at
http://www.taylorandfrancis.com**

**and the CRC Press Web site at
http://www.crcpress.com**

In my mother's memory

Contents

Preface

This book aims at providing an introduction to partial differential equations (PDEs) and expects to serve as a textbook for the young graduate students of theoretical physics and applied mathematics who have had an initial training in introductory calculus and ordinary differential equations (ODEs) and an elementary grasp of classical mechanics. The presentation of the book does not always follow the conventional practice of classifying the PDE and then taking up the treatment of the representative Laplace equation, wave equation and heat conduction equation one by one. On the contrary, certain basic features of these equations are presented in the introductory chapter itself with an aim to provide an appropriate formulation of mathematical methods associated with them as and when they emerge from the class of PDE they belong to.

Numerous examples have been worked out in the book which aim to illustrate the key mathematical concepts involved and serve as a means to improve the problemsolving skills of the students. We have purposefully kept the complexities of the mathematical structure to a minimum, often at the expense of general and abstract formulation, but put greater stress on the application-side of the subject. We believe that the book will provide a systematic and comprehensive coverage of the basic theory of PDEs.

The book is organized into six chapters and three appendices. Chapter 1 contains a general background of notations and preliminaries required that would enable one to follow the rest of the book. In particular, it contains a discussion of the first order PDEs and the different types of equations that one often encounters in practice. The idea of characteristics is presented and also second and higher order PDEs are introduced. We also comment briefly on the Cauchy problem and touch upon the classification of a second order partial differential equation in two variables.

Chapter 2 shows how a PDE can be reduced to a normal or canonical form and the utility in doing it. We give some insights from the classical mechanics as well. We also discuss the construction of the adjoint and self-adjoint operators.

The elliptic form of a partial differential equation is discussed in Chapter 3. Topics like method of separation of variables for the two-dimensional plane polar coordinates, three-dimensional spherical polar coordinates and cylindrical polar coordinates, harmonic functions and their properties, maximum-minimum principle, existence, uniqueness and stability of solutions, normally directed distribution of doublets, Green's equivalent layer theorem, generation of Green's function, Dirichlet's problem of a circle, sphere and half-space are considered.

PDEs of hyperbolic type are taken up in Chapter 4 which begins with the D'Alembert's solution of the linear second order wave equation. The more general Riemann's method is introduced next. The role of Riemann function is pointed out to help us solve the PDE in the form of a quadrature. Solutions by the method of separation of variables is illustrated for the three-dimensional wave equation both for the spherical polar and cylindrical coordinates. This chapter also consists of detailed discussions of the initial value problems related to the three-dimensional and two-dimensional wave equations. We give here the basic derivation of the Poisson/Kirchoff solution for the three-dimensional homogeneous equation supplemented by a set of inhomogeneous conditions and then focus on the solution of the inhomogeneous wave equation by means of the superposition principle, which gives us the Poisson formula. Hadamard's method of descent is employed to solve completely the two-dimensional counterpart.

Parabolic equations in Chapter 5 is our next topic of inquiry. This chapter covers the Cauchy problem for the heat equation wherein we discuss the uniqueness criterion of the solution, the method of separation of variables for the Cartesian coordinates, spherical polar and cylindrical polar coordinates, and derive the fundamental solution and give a formulation of the Green's function.

Chapter 6 is concerned with solving different types of PDEs by the integral transform method focusing on the use of Fourier transform and Laplace transforms only. Some problems are worked out for their asymptotic nature.

Finally in the three appendices we address respectively some important issues of the delta function, the Fourier transform and Laplace transform. I wish to remark that we added a summary at the end of each chapter for the lay reader to have a quick look at the materials that have been covered and also appended a reasonable collection of homework problems relevant to the chapter. In addition, we solved in each chapter a wide range of problems to clarify the basic ideas involved. It is believed that these will help the readers in furthering the understanding of the subject.

Acknowledgments

I would like to express my gratitude to Prof. Rupamanjari Ghosh, Vice Chancellor, Shiv Nadar University for her steadfast support without which it would not have been possible to complete the book. I also appreciate the continual encouragement from Prof. Sankar Dhar, Head, Department of Physics, Shiv Nadar University during the preparation of this book.

It gives me great pleasure to acknowledge the kind help that I received, time and time again, from Prof. Kalipada Das, Department of Applied Mathematics, University of Calcutta that saw me through this book. I am very grateful to him. I also recall the insightful comments of the late Mithil Ranjan Gupta of the same department that helped my understanding enormously. I thank Dr. Santosh Singh, Department of Mathematics, Shiv Nadar University for many helpful conversations and countless acts of support, Dr. Santosh Kumar, Department of Physics, Shiv Nadar University for frequent help, and Prof. Somnath Sarkar, Department of Electronics, University of Calcutta for his interest in the development of this book. I owe gratitude to my students Mr. Supratim Das and Ms. Debanjana Bose for organizing the first few chapters of my lecture notes, and Mr. Yogesh Yadav for carefully reading certain portions and offering valuable suggestions. I would also like to thank Ms. Aastha Sharma, Commissioning Editor, CRC Press, Taylor & Francis Group, and her editorial team, especially Ms. Shikha Garg for extending support toward finalizing the draft of the book.

Finally, I thank my wife Minakshi, and daughter Basabi for always urging me to turn the idea of writing a book on partial differential equations into a reality.

Bijan Kumar Bagchi
Department of Physics,
Shiv Nadar University,
Greater Noida

Author

Bijan Kumar Bagchi received his B.Sc., M.Sc., and Ph.D. from the University of Calcutta. He has a variety of research interests and involvements ranging from spectral problems in quantum mechanics to exactly solvable models, supersymmetric quantum mechanics, parity-time symmetry and related non-Hermitian phenomenology, nonlinear dynamics, integrable models and high energy phenomenology.

Dr. Bagchi has published more than 150 research articles in refereed journals and held a number of international visiting positions. He is the author of the books entitled *Advanced Classical Mechanics* and *Supersymmetry in Quantum and Classical Mechanics* both published by CRC Press, respectively, in 2017 and 2000. He was formerly a professor in applied mathematics at the University of Calcutta, and is currently a professor in the Department of Physics at Shiv Nadar University.

Chapter 1

Preliminary concepts and background material

An ordinary differential equation (ODE) is an expression that involves one or more functions defined with respect to one independent variable along with the derivatives of such functions while a partial differential equation (PDE) is concerned with a relationship involving an unknown function of several independent variables and their partial derivatives with respect to these variables. The utility of an ODE is limited because a physical system very often depends on more than one independent variable. On the other hand, with the presence of a multitude of independent variables (at least two), PDEs address a far wider audience of various branches of mathematical and theoretical physics.

PDEs find numerous applications for different ranges of physical phenomena. Some commonplace examples are the Navier-Stokes equation of fluid mechanics, Maxwell's equations of electrodynamics, problems of population dynamics and dynamical systems. PDEs are also encountered in various other disciplines such as plasma dynamics, quantum mechanics, mathematical biology, computational chemistry, areas dealing with differential geometry, complex analysis, symplectic geometry and algebraic geometry apart from other pursuits seeking analytical or computational results arising in problems of mathematical methods.

The study of differential equations can be traced back to the basic works of Issac Newton (1642-1727) and Gottfried Leibniz (1646-1716). While Newton dealt with what he called *fluxional equations*, Leibniz made an extensive study of the first order equations like $a(x,y)\frac{\partial u}{\partial x} + b(x,y)\frac{\partial u}{\partial y} = c(x,y)$ and introduced shorthands like δ to denote partial derivatives $\frac{\partial}{\partial x}$ or $\frac{\partial}{\partial y}$. Wide ranging applications of differential equations to classical mechanics, hydrodynamics, astronomy and related areas were subsequently pioneered by the Bernoulli brothers Jakob (1654-1705) and Johann (1667-1748), Johann's illustrious son Daniel (1700-1782), Leonhard Euler (1707-1783), Jean D'Alembert (1717-1783), Joseph Lagrange (1736-1813) and Pierre Simon de Laplace (1749-1827). Certain aspects of mathematical rigour were exploited by Augustin Cauchy (1789-1859) who proved a number of existence theorems while other important contributions came from Johann Pfaff (1765-1825), Joseph Fourier (1768-1830), Siméon Poisson (1781-1840), Carl Jacobi (1804-1851) and Sophus Lie (1842-1899).

1.1 Notations and definitions

We shall normally adopt the following notations :

Independent variables : x, y, z, ξ, η, ...

Dependent variables : ϕ, ψ, u, v, w, ...

Partial derivatives : $\phi_x(\equiv \frac{\partial\phi}{\partial x})$, $\phi_y(\equiv \frac{\partial\phi}{\partial y}$, $\phi_{xx}(\equiv \frac{\partial^2\phi}{\partial x^2})$, $\phi_{xy}(\equiv \frac{\partial^2\phi}{\partial x\partial y})$, $\phi_{yy}(\equiv \frac{\partial^2\phi}{\partial y^2})$, ...

The general form of a partial differential equation involving two independent variables, x and y, is given by

$$F(x,y,\phi_x,\phi_y,\phi_{xx},\phi_{yy},\phi_{xy},\phi_{xxx},\phi_{yyy},...) = 0, \quad x,y \in \Omega \qquad (1.1)$$

where Ω is some specified domain, F is a functional of the arguments indicated and ϕ is an arbitrary function of (x, y). By the order of a PDE we mean the order of the highest derivative appearing in (1.1). A solution of (1.1) corresponds to the function $\phi(x, y)$ obeying (1.1) for all values of x and y.

If we are dealing with n independent variables, $x_1, x_2, ..., x_n$, the domain Ω will refer to an $n-$ dimensional space containing an $(n - 1)$-dimensional hypersurface. A hypersurface of dimension $(n - 1)$ is given by an equation of the form $x_1^2 + x_2^2 + ... + x_n^2 = 1$ residing in an n-dimensional Euclidean space. For instance, in an Euclidean space of dimension two, a hypersurface is a plane curve while in an Euclidean space of dimension three, a hypersurface is a surface.

Some simple examples of a PDE are

Example 1.1 $\phi_x - b\phi_y = 0 : b$ is a constant: the PDE is first order in x and y.

Example 1.2 $\phi_{xx} + \phi_y = 0 :$ the PDE is second order in x but first order in y.

Example 1.3 $\phi_x + \phi_{yy} = 0 :$ the PDE is first order in x but second order in y.

In this book we will be interested mostly in the linear forms of the PDE. Adopting an operator notation a linear PDE can be put in the form

$$L\phi(x, y) = \rho(x, y) \tag{1.2}$$

where L denotes a suitable operator, all terms involving ϕ and its derivatives are grouped in the left side while the term $\rho(x, y)$ in the right side, which acts as a source term, is assumed to be known. Formally, a linear PDE is the one which is linear in the dependent variable and all of its partial derivatives. Linearity is governed by the following properties of L

$$(i) L(\phi + \psi) = L\phi + L\psi$$
$$(ii) L(c\phi) = cL\phi$$

for arbitrary functions ϕ, ψ and a constant c.

Example 1.4

The heat conduction equation

$$\phi_t - k\phi_{xx} = 0, \quad t > 0 \quad 0 < x < l$$

where k is a constant, is clearly linear.

Example 1.5

The PDE

$$\phi\phi_t + \phi = \sin x$$

is nonlinear because of the product term $\phi\phi_t$.

Taking L to be a linear operator, if $\rho(x, y) = 0$ in (1.2), the PDE is classified as a linear homogeneous equation. The latter then reads simply

$$L\phi(x, y) = 0 \qquad (1.3)$$

When this is not the case i.e. if $\rho(x, y) \neq 0$, the PDE is said to be an inhomogeneous (or non-homogeneous) equation. We therefore observe that every inhomogeneous equation has a corresponding homogeneous counterpart. In Example (1.5), $\sin x$ stands for the inhomogeneous term. However its presence does not alter the linear or nonlinear character of the equation. A general solution of the homogeneous equation when superposed with the particular solution of the inhomogeneous equation produces the general solution of the inhomogeneous equation.

For a first order linear equation, $L\phi$ reads

$$L\phi(x, y) \equiv a(x, y)\phi_x + b(x, y)\phi_y + c(x, y)\phi$$

where the coefficients a, b, c are functions of x and y while for a second order linear equation, $L\phi$ can be projected in its general form

$$\begin{aligned} L\phi(x, y) &\equiv A(x, y)\phi_{xx} + 2B(x, y)\phi_{xy} + C(x, y)\phi_{yy} \\ &\quad + D(x, y)\phi_x + E(x, y)\phi_y + F(x, y)\phi \end{aligned}$$

where the coefficients A, B, C, D, E and F are functions of x and y.

If the PDE is not linear then it is a nonlinear PDE. Note however that if an equation is linear in the highest derivatives of ϕ then such an equation is called a quasi-linear equation. A couple of typical examples with respect to two independent variables x and y are the first order equation

$$(1 + \phi^2)\phi_x + \phi_y = x^2$$

and the second order equation

$$\phi_x\phi_{xx} + \phi_y\phi_{yy} = \phi^3$$

On the other hand, if the coefficients of the highest order derivatives of a quasi-linear equation are functions only of the independent variables then such an equation is called an almost linear, or half-linear, or simply a semi-linear PDE. An appropriate example of a semi-linear equation is

$$x^2\phi_{xx} + 4xy\phi_{yy} + \phi\phi_x + \phi^2 = 0$$

where x and y are the independent variables.

To extract a solution of a PDE one needs to have some knowledge of the associated initial or boundary conditions. If the PDE is equipped with a condition at an initial time, say at $t = 0$, for a certain portion of the region under consideration, it is referred to as an initial condition. A condition that holds on any other curve for all times in that region is called a boundary condition. Sometimes we encounter an eigenvalue problem too. An appropriate set of boundary or initial condition is expected to yield a unique and stable solution.

Let us explain all this by means of some examples from the theory of ODE. Suppose that the temperature T of a given material is given to be T_0 at some initial time $t = 0$. We can solve such an initial value problem by assigning for the temperature variation a law of cooling that gives the rate of cooling as directly proportional to the temperature difference between the object and its surroundings. We assume that the temperature of the surroundings is zero. Then the guiding differential equation takes the form

$$
\begin{aligned}
T'(t) &= -kT(t) \qquad t > 0 \\
T(0) &= T_0
\end{aligned}
$$

where k is a constant of proportionality. The solution turns out to be exponentially damping

$$T(t) = T_0 e^{-kt} \qquad t \geq 0$$

Note that the complementary case corresponds to the differential equation $T'(t) = kT(t)$ which provides the character of the solution as an exponentially expanding one.

Next, consider the simple case of fitting a straight line as given by the standard equation $y(x) = mx + c$ which connects two points, say x_1 and x_2 in a certain given plane. This criterion determines both the unknowns m and c enabling us to express the straight line equation as

$$y(x) = \left(\frac{y_2 - y_1}{x_2 - x_1}\right)x + \left(\frac{y_1 x_2 - y_2 x_1}{x_2 - x_1}\right), \qquad x_1 < x < x_2$$

where y_1 and y_2 denote the y values for $x = x_1$ and x_2 respectively.

This is an example of a boundary value problem corresponding to a given prescription on the function $y(x)$ in the domain $[x_1, x_2]$. Mathematically we can formulate the whole scheme in the form of a second order ODE namely,

$$y''(x) = 0 \qquad x_1 < x < x_2$$

where the primes correspond to derivatives with respect to the independent variable x.

Finally, we can think of oscillating functions such as $\phi(x) = \sin kx$, where k is a constant, which emerge as solutions of the differential equation $\phi''(x) = -k^2\phi(x)$, $-\infty < x < \infty$. As is evident, we have an eigenvalue problem which can be solved if suitable initial inputs are provided.

1.2   Generating a PDE

We can generate PDEs in different ways. We discuss here two possibilities.

(a) Eliminating arbitrary constants from a given relation

Consider a two-dimensional quadratic equation

$$(x - \alpha)^2 + (y - \beta)^2 + \phi^2 = 1$$

where ϕ is a function of two independent variables x and y and α, β are two constants. Partially differentiating both sides with respect to x gives

$$(x - \alpha) + \phi\phi_x = 0$$

while partially differentiating both sides with respect to y results in

$$(y - \beta) + \phi\phi_y = 0$$

If we eliminate the constants α and β from the last two equations we obtain a first order PDE

$$\phi^2[1 + (\phi_x)^2 + (\phi_y)^2] = 1$$

Next, consider a somewhat different type of quadratic equation namely

$$\frac{x^2}{a^2} + \frac{y^2}{b^2} + \frac{\phi^2}{c^2} = 1$$

where ϕ is a function of two independent variables x and y but the equation involves three constants a, b and c.

Partially differentiating both sides of this equation with respect to x gives

$$\frac{2x}{a^2} + \frac{2\phi}{c^2}\phi_x = 0 \quad \Rightarrow \quad \frac{c^2}{a^2} = -\frac{\phi}{x}\phi_x$$

If partially differentiated again with respect to x we get

$$\frac{2}{a^2} + \frac{2}{c^2}(\phi_x)^2\phi_{xx} = 0 \quad \Rightarrow \quad \frac{c^2}{a^2} = -(\phi_x)^2 - \phi\phi_{xx}$$

Eliminating $\frac{c^2}{a^2}$ from the last two equations leads to a second order PDE in terms of the independent variable x

$$x\phi\phi_{xx} + x(\phi_x)^2 - \phi\phi_x = 0$$

In a similar way if we carry out partial differentiations with respect to the variable y twice we get a second order PDE in terms of the independent variable y

$$y\phi\phi_{yy} + y(\phi_y)^2 - \phi\phi_y = 0$$

What do we infer from the two examples considered?

If the number of constants to be eliminated equals the number of independent variables, as was the case in the first example, then we run into a first order PDE. However, if the number of constants to be eliminated is greater than the number of independent variables, as we saw in the case of the second example, then we obtain an equation of second or higher order.

(b) Elimination of an arbitrary function

Let us now deal with two functions $F = F(x, y, \phi)$, $G = G(x, y, \phi)$ and assume a certain connection χ existing between them

$$\chi(F, G) = 0$$

The above relation, on taking partial derivatives with respect to x and y, yields respectively

$$\chi_F(F_x + F_\phi\phi_x) + \chi_G(G_x + G_\phi\phi_x) = 0$$

$$\chi_F(F_y + F_\phi\phi_y) + \chi_G(G_y + G_\phi\phi_y) = 0$$

It is an easy task to get rid of χ_F and χ_G which yields

$$(F_x + F_\phi\phi_x)(G_y + G_\phi\phi_y) - (F_y + F_\phi\phi_y)(G_x + G_\phi\phi_x) = 0$$

The above expression simplifies because the term $F_\phi\phi_x G_\phi\phi_y$ cancels out and we are left with the form

$$(F_\phi G_y - F_y G_\phi)\phi_x + (F_x G_\phi - F_\phi G_y)\phi_y = F_y G_x - F_x G_y$$

This gives the Lagrange's form of a first order quasi-linear PDE

$$a(x, y, \phi)\phi_x + b(x, y, \phi)\phi_y = c(x, y, \phi)$$

where the coefficients a and b do not involve the arbitrary connecting function χ.

1.3   First order PDE and the concept of characteristics

We inquire into solving a linear first order PDE of the form

$$a(x,y)\phi_x + b(x,y)\phi_y = 0$$

where we regard $\phi = \phi(x,y)$ as a surface which is generated by the family of curves C_1, C_2, ...

Let us parametrize a typical curve C_i in terms of a parameter s as given by the following pair of differential equations

$$\frac{dx}{ds} = a(x,y) \qquad\qquad \frac{dy}{ds} = b(x,y)$$

Then the derivative of ϕ with respect to s reads

$$\frac{d\phi}{ds} = \phi_x \frac{dx}{ds} + \phi_y \frac{dy}{ds} = a(x,y)\phi_x + b(x,y)\phi_y = 0;$$

In other words, along C_i, ϕ is a function of s only. Its solution is $\phi = k = $ constant and so along C_i, ϕ is a constant. C_i is referred to as a characteristic curve.

The equations of the characteristics are obtained by solving the subsidiary equations:

$$\frac{dx}{a} = \frac{dy}{b}$$

in which the parameter s does not appear. Since the solution involves one arbitrary constant, the characteristics, at the first order level, generate a one-parameter family of curves.

Example 1.6

Consider the first order PDE for $\phi(x,y)$

$$\phi_x + x\phi_y = 0$$

subject to $\phi(0,y) = 1 + y^2$ where $0 < y < 2$.

The pair of characteristics are readily seen to be given by the equations

$$\frac{dx}{ds} = 1 \qquad\qquad \frac{dy}{ds} = x$$

Elimination of ds gives

$$x^2 - 2y = k, \qquad k = \text{constant}$$

implying that the resulting characteristics are parabolas.

On each of the parabolas, ϕ is constant. The form of ϕ in the region R bounded by the characteristics passing through the origin $(0,0)$ and $(0,2)$ can be determined as follows.

Let $(\bar{x}, \bar{y})$ be any point in R. The equation of the characteristic passing through the point $(\bar{x}, \bar{y})$ is given by

$$x^2 - 2y = \bar{x}^2 - 2\bar{y} = k$$

On the above parabola, ϕ must be a constant say, λ

$$\phi = \lambda$$

This parabola will intersect the line segment between $(0,0)$ and $(0,2)$ at the point where $x = 0$ and the corresponding y coordinate y_0 given by

$$0^2 - 2y_0 = \bar{x}^2 - 2\bar{y} \Rightarrow y_0 = \frac{1}{2}(2\bar{y} - \bar{x}^2)$$

Comparing with the given condition $\phi(0, y) = 1 + y^2$ we see that

$$\lambda = 1 + \frac{1}{4}(2\bar{y} - \bar{x}^2)^2$$

Since $(\bar{x}, \bar{y})$ is any point in R, the solution of ϕ in R is

$$\phi = 1 + \frac{1}{4}(2y - x^2)^2$$

Here the stringent requirement has been that the line segment connecting $(0,0)$ and $(0,2)$ is not a part of any characteristic as shown in Figure 1.1.

1.4 Quasi-linear first order equation: Method of characteristics

(a) Lagrange's method of seeking a general solution

Let us address Lagrange's form of the quasi-linear first order PDE

$$a(x, y, \phi)\phi_x + b(x, y, \phi)\phi_y = c(x, y, \phi) \tag{1.4}$$

where a, b, c are functions of x, y and ϕ. Note that the functions a, b, c do not involve the derivatives of ϕ.

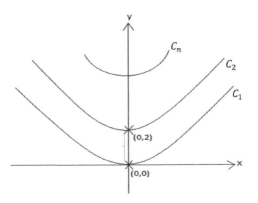

FIGURE 1.1: The parabolas through the points $(0,0)$ and $(0,2)$.

In what follows we introduce Lagrange's method to demonstrate that the general solution to (1.4) is of the type

$$\chi(F,G) = 0$$

where χ is an arbitrary function of F and G with $F(x,y,\phi) = k_1$, k_1 is a constant and $G(x,y,\phi) = k_2$, k_2 is a constant, corresponding to two independent solutions of the subsidiary equations

$$\frac{dx}{a(x,y,\phi)} = \frac{dy}{b(x,y,\phi)} = \frac{d\phi}{c(x,y,\phi)} \tag{1.5}$$

Proof:

Partially differentiating $\chi(F,G)$ with respect to x and y gives respectively

$$\chi_x + \chi_\phi \phi_x = 0, \quad \chi_y + \chi_\phi \phi_y = 0 \tag{1.6}$$

The above equations imply that

$$\phi_x = -\frac{\chi_x}{\chi_\phi}, \quad \phi_y = -\frac{\chi_y}{\chi_\phi} \quad \text{(we let } \chi_\phi \neq 0) \tag{1.7}$$

As a result the Lagrange's equation is converted to the form

$$a\chi_x + b\chi_y = c\chi_\phi \tag{1.8}$$

On the other hand, the subsidiary equations associated with the above form read

$$\frac{dy}{dx} = \frac{b}{a}, \quad \frac{d\phi}{dx} = \frac{c}{a} \tag{1.9}$$

To prove that $F(x, y, \phi) = k_1$ and $G(x, y, \phi) = k_2$ are solutions of the Lagrange's equation we take the differential of F and G to write down

$$dF \equiv F_x dx + F_y dy + F_\phi d\phi = 0, \quad dG \equiv G_x dx + G_y dy + G_\phi d\phi = 0 \tag{1.10}$$

Due to the subsidiary equations (1.9) these equations assume the forms

$$aF_x + bF_y + cF_\phi = 0, \quad aG_x + bG_y + cG_\phi = 0 \tag{1.11}$$

showing both F and G to be solutions of Lagrange's equation. The general solution is thus expressible in the functional form

$$\chi(F, G) = 0 \tag{1.12}$$

with χ being an arbitrary function.

(b) Integral lines and integral surfaces

Let $\phi(x, y)$ be a solution of (1.4). We can represent the latter equation in the form

$$(a, b, c) \cdot (\phi_x, \phi_y, -1) = 0 \tag{1.13}$$

Since $(\phi_x, \phi_y, -1)$ is normal to the surface $S = [x, y, \phi(x, y)]$, it follows that the vector field $V = (a, b, c)$ lies in the tangent plane to S. This is indeed a feature of the solution $\phi(x, y)$ we are looking for. The surface for which the vector field V lies in its tangent plane, at each point in the surface, is called an integral surface. In fact any point on the integral surface has the form $(x, y, \phi(x, y))$.

We now address the Cauchy problem associated with the Lagrange's equation (1.4). It is concerned with the determination of its unique solution $\phi(x, y)$ such that it takes given values on a regular arc[1] γ in the xy-plane.

---

[1] A plane arc is defined as a collection of points which can be parametrized by the coordinates $x = f(t)$ and $y = g(t)$ with the parameter t belonging to some closed interval

Let the parametric form of the curve γ be

$$\gamma: \quad x = f(t), \quad y = g(t) \tag{1.14}$$

We can look upon γ as a projection in the xy-plane of the curve Γ- defined in a three-dimensional $xy\phi$-space by

$$\Gamma: \quad x = f(t), \quad y = g(t), \quad h = h(t) \tag{1.15}$$

where

$$h(t) = \phi(f(t), g(t)) \tag{1.16}$$

The coefficient functions in (1.4) are now all expressible in terms of a single parameter t and we assume them to be analytic: in other words, differentiable sufficient numbers of times. The functions a, b and c are also assumed to be differentiable sufficient numbers of times.

At the point (x_0, y_0). equation (1.4) takes the form

$$a_0(\phi_x)_0 + b_0(\phi_y)_0 = c_0 \tag{1.17}$$

where the suffix (0) indicates the value at (x_0, y_0). Next, differentiating (1.16) with respect to t and evaluating the resulting expression at (x_0, y_0) yields

$$(\phi_x)_0 f'(t_0) + (\phi_y)_0 g'(t_0) = h'(t_0) \tag{1.18}$$

where the prime stands for differentiation with respect to t.

It is clear that a unique solution of $(\phi_x)_0$ and $(\phi_y)_0$ would emerge from (1.17) and (1.18) provided the discriminant

$$\Delta \equiv a_0 g'(t_0) - b_0 f'(t_o) \neq 0 \tag{1.19}$$

The condition (1.19) is crucial for the solvability of the Cauchy problem.

We also see that if we eliminate $(\phi_x)_0$ from (1.17) and (1.18) we obtain

$$[a_0 g'(t_o) - b_0 f'(t_o)](\phi_y)_0 = a_0 h'(t_0) - c_0 f'(t_o) \tag{1.20}$$

Thus $\Delta = 0$ is equivalent to

$$\Delta = 0: \quad a_0 h'(t_0) - c_0 f'(t_o) = 0 \tag{1.21}$$

as well.

$[a, b]$, $f(t)$ and $g(t)$ being continuous functions of t. A simple arc is one in which no point on it corresponds to two different values of t. If $f(t)$ and $g(t)$ are also continuously differentiable (with the derivatives being one-sided at the end points) then the simple arc is termed a regular arc.

We therefore have for the vanishing of the discriminant the following criterion

$$\Delta = 0 : \quad \frac{f'(t_0)}{a_0} = \frac{g'(t_0)}{b_0} = \frac{h'(t_0)}{c_0} \tag{1.22}$$

which from (1.15) points to the subsidiary or auxiliary equations

$$\Delta = 0 : \quad \frac{dx}{a} = \frac{dy}{b} = \frac{d\phi}{c} \tag{1.23}$$

These auxiliary equations amount to a set of ODEs.

We can state equivalently that the curve $\Gamma = [x(t), y(t), h(t)]$ satisfies the ODEs namely,

$$\frac{dx}{dt} = a(x(t), y(t)), \quad \frac{dy}{dt} = b(x(t), y(t)), \quad \frac{d\phi}{dt} = c(x(t), y(t)) \tag{1.24}$$

We call this curve an integral curve for the vector field having the components (a, b, c). The integral curves are known as the characteristic curves for (1.4). The notable feature of a characteristic curve Γ is that it would support infinitely many solutions for the Cauchy problem. A unique solution of the Cauchy problem will only exist as long as Γ is non-characteristic.

The characteristic curves are obtained by solving (1.24). The projected curves on the xy- plane is determined by solving the differential equation $\frac{dy}{dx} = \frac{b(x,y)}{a(x,y)}$. Union of the characteristic curves gives the integral surface. Formally, it corresponds to assembling the surface $\phi = \phi(x, y)$ through each point of the characteristics. The following examples will make the point clear.

Example 1.7

Find the integral lines and integral surface of the one-dimensional transport equation

$$\phi_\tau + c\phi_x = 0$$

where c is a constant. The initial condition is provided by $\phi(x, 0) = f(x)$.

Here the corresponding equations to (1.24) are

$$\frac{d\tau}{dt} = 1, \quad \frac{dx}{dt} = c, \quad \frac{d\phi}{dt} = 0$$

Their solutions are rather simple

$$x(t) = ct + \lambda, \quad \tau(t) = t + \mu, \quad z(t) = \nu$$

where λ, μ and ν are constants of integration. From the first two equations we can get rid of the parameter t to have $x - c\tau = k$, $k = \lambda - c\mu$, which along with $z(t) = \nu$ give the integral curves as straight lines in the three-dimensional $xy\phi$-space. The integral surface is a collection of these lines. Note that along these lines the solution $\phi(x, \tau)$ is constant (which can be checked by taking the τ- derivative: $\frac{d\phi}{d\tau} = c\phi_x + \phi_\tau = 0$) and we may set $\phi(x, \tau) = f(x - c\tau)$ which gives the general solution.

Example 1.8

Discuss the general solution of the quasi-linear equation

$$(y + x\phi)\phi_x + (x + y\phi)\phi_y = \phi^2 - 1$$

Find the integral surface through the parabola $x = t, y = 1, \phi = t^2$.
Here the equations corresponding to (1.24) are

$$\frac{dx}{y + x\phi} = \frac{dy}{x + y\phi} = \frac{d\phi}{\phi^2 - 1}$$

By summing and subtracting the first two equalities and equating with the third we can easily deduce

$$\frac{d(x + y)}{x + y} = \frac{d\phi}{\phi - 1}$$

and

$$\frac{d(x - y)}{x - y} = \frac{d\phi}{\phi + 1}$$

Solving the above two differential equations we find for the respective equations

$$\phi = 1 + \lambda(x + y), \quad \phi = -1 + \mu(x - y)$$

where λ and μ are two arbitrary constants of integration.
The general solution is therefore given by

$$F(\frac{\phi - 1}{x + y}, \frac{\phi + 1}{x - y}) = 0$$

where F is an arbitrary function of its arguments.

To get the integral surface we note that for the given parabola the constant λ and μ turn out to be

$$\lambda = t - 1, \quad \mu = \frac{t^2 + 1}{t - 1}$$

which on elimination of t implies that λ and μ are subjected to the constraint

$$\lambda\mu = (1 + \lambda)^2 + 1$$

Hence using the relations for λ and μ we find for the integral surface the equation

$$\frac{(\phi - 1)(\phi + 1)}{(x + y)(x - y)} = (\frac{\phi - 1}{x + y} + 1)^2 + 1$$

That is

$$(\phi + x + y - 1)^2 + (x + y)^2 = \frac{(x + y)(\phi^2 - 1)}{x - y}$$

Example 1.9

Find the integral surface of the following quasi-linear equation

$$(y - \phi)\frac{\partial \phi}{\partial x} + (\phi - x)\frac{\partial \phi}{\partial y} = x - y$$

which goes through the curve $\phi = 0, xy = 1$ and through the circle $x + y + \phi = 0, x^2 + y^2 + \phi^2 = a^2$.

Here the equations corresponding to (1.24) are

$$\frac{dx}{y - \phi} = \frac{dy}{\phi - x} = \frac{d\phi}{x - y}$$

By summing the numerator and denominator each fraction turns out to be

$$= \frac{dx + dy + d\phi}{0}$$

On the other hand, if we multiply the numerator and denominator of the first fraction by x, the second fraction by y and the third fraction by ϕ and sum them then each fraction takes the form

$$= \frac{xdx + ydy + \phi d\phi}{0}$$

It at once follows that

$$dx + dy + d\phi = 0 \Rightarrow x + y + \phi = \lambda$$

and

$$x dx + y dy + \phi d\phi = 0 \Rightarrow x^2 + y^2 + \phi^2 = \mu$$

where λ and μ are two arbitrary constants of integration.

Therefore the general solution is

$$F(x + y + \phi, x^2 + y^2 + \phi^2) = 0$$

where F is an arbitrary function of its arguments.

On the given curve $\phi = 0$ and $xy = 1$, let us parametrize the latter by $x = t, y = \frac{1}{t}$. Then the two constraints above read

$$t + \frac{1}{t} = \lambda, \quad t^2 + \frac{1}{t^2} = \mu$$

and we have the relation

$$\mu = \lambda^2 - 2$$

This implies

$$x^2 + y^2 + \phi^2 = (x + y + \phi)^2 - 2$$

On simplification the integral surface is

$$xy + y\phi + x\phi = 1$$

Sometimes solving in the parametric form proves tedious. Consider the following example.

Example 1.10

Find the general solution and the characteristic curves of the following PDE

$$\phi\phi_y + 2y\phi_x = 2y\phi^2$$

Here the corresponding equations to (1.4) are

$$\frac{dy}{dt} = \phi, \quad \frac{dx}{dt} = 2y, \quad \frac{d\phi}{dt} = 2y\phi^2$$

Here if one tries to eliminate the parameter t from the first and third equation one runs into a nonlinear differential equation $\ddot{y} = 2y\dot{y}^2$, where the dot refers to a derivative with respect to t. We therefore take an alternative approach. We see that the first two equalities give $\frac{dy}{dx} = \frac{\phi}{2y}$ while the first and third equalities lead to $\frac{d\phi}{dy} = 2y\phi$. While the solution of the second differential equation gives $\phi = \lambda e^{y^2}$, the first one gives $(x\phi + 1)e^{-y^2} = \mu$ when the last solution is used. Note that λ and μ are arbitrary constants.

The general solution is thus

$$F((x\phi + 1)e^{-y^2}, \phi e^{-y^2}) = 0$$

implying

$$\phi(x, y) = e^{y^2} f((x\phi + 1)e^{-y^2})$$

where f is an arbitrary function and the characteristics are $\phi e^{-y^2} = \text{constant}$ and $(x\phi + 1)e^{-y^2} = \text{constant}$.

We conclude this section by making a few remarks on the Taylor series expansion of the function $\phi(x, y)$. Expanding about a point $t = t_0$ corresponding to which $x_0 = f(t_0)$ and $y_0 = g(t_0)$ we can write

$$\phi(x, y) = \phi_0 + [(x - x_0)(\phi_x)_0 + (y - y_0)(\phi_y)_0] +$$

$$+ \frac{1}{2!}[(x - x_0)^2(\phi_{xx})_0 + 2(x - x_0)(y - y_0)(\phi_{xy})_0 + (y - y_0)^2(\phi_{yy})_0] + \dots$$

$$(1.25)$$

If the condition (1.19) holds, in other words, Γ is non-characteristic at the point (x_0, y_0), then all the terms in the right side (1.25) corresponding to the first and higher order derivatives of ϕ at $t = t_0$ can be uniquely determined from the given values on γ.

In fact on partially differentiating with respect to x we have at $t = t_0$

$$(\phi_{xx})_0 f'(t_0) + (\phi_{xy})_0 g'(t_0) = \phi'_x(t_0) \tag{1.26}$$

where the prime denotes a derivative with respect to t.

Next, differentiating (1.4) partially with respect to x

$$a\phi_{xx} + b\phi_{xy} = Q(x, y, \phi, \phi_x, \phi_y) \tag{1.27}$$

where Q stands for

$$Q(x, y, \phi, \phi_x, \phi_y) = c_x + c_\phi \phi_x - (a_x + a_\phi \phi_x)u_x - (b_x + b_\phi \phi_x)\phi_y \tag{1.28}$$

At $t = t_0$, (1.27) yields

$$a_0(\phi_{xx})_0 + b_0(\phi_{xy})_0 = Q(x_0, y_0, \phi_0, (\phi_x)_0, (\phi_y)_0) \tag{1.29}$$

Since Γ is non-characteristic at $t = t_0$ and in its neighbourhood, we can uniquely determine the quantities $(\phi_{xx})_0$ and $(\phi_{xy})_0$ from (1.26) and (1.29). Note that $(\phi_{yy})_0$ is uniquely determined by taking a derivative partially with respect to y instead of x and derive analogous equations to (1.27) and (1.29) containing the quantity $(\phi_{yy})_0$. Higher derivatives at $t = t_0$ are estimated by proceeding in a similar manner. Thus a unique solution of (1.4) can be obtained by forming a Taylor expansion for $\phi(x, y)$ which assumes given values on a non-characteristic curve γ.

1.5   Second order PDEs

We give now some typical examples of PDE which we often come across in problems of mathematical physics. These are

(1) Poisson's equation and Laplace's equation (electrostatics, Newtonian gravity)

$$\nabla^2 \phi(\mathbf{r}, t) = -(4\pi)\rho$$

where ρ is a charge distribution indicating the presence of a source term. With $\rho \neq 0$, Poisson's equation is an inhomogeneous PDE.

When $\rho = 0$, we get the Laplace's equation

$$\nabla^2 \phi(\mathbf{r}, t) = 0$$

Laplace's equation appears in problems of electrostatics, Newtonian gravity and hydrodynamics.

(2) Wave equation (acoustics, electrodynamics)

$$\nabla^2 \phi(\mathbf{r}, t) - \frac{1}{c^2} \phi_{tt}(\mathbf{r}, t) = 0$$

The vibration of a string of constant density under constant tension is guided by an equation of the above form. In acoustics, c stands for the velocity of sound while in electrodynamics, c refers to the velocity of light.

(3) Helmholtz's equation (vibrating string problem)

$$(\nabla^2 + k^2)\phi(\mathbf{r}, t) = 0$$

where k is a constant. It appears straightforwardly when the wave equation is subjected to a Fourier transform with respect to the time variable. Separation of variable when applied to the wave equation also results in Helmholtz's equation to which we will come later.

(4) Heat conduction equation (equalization of energy process)

$$\nabla^2 \phi(\mathbf{r}, t) - \frac{1}{k} \phi_t(\mathbf{r}, t) = 0$$

where k is called the temperature conductivity. It has relevance in the equalization of energy process.

(5) Schrödinger equation (non-relativistic quantum mechanics)

$$i\hbar \frac{\partial \psi(\mathbf{r}, t)}{\partial t} = [-\frac{\hbar^2}{2\mu} \nabla^2 + V(\mathbf{r}, t)]\psi(\mathbf{r}, t)$$

where $\hbar(= \frac{h}{2\pi})$ is the reduced Planck's constant, ψ is the wave function that guides the quantum system and V is the potential energy. Schrödinger equation provides a description of the dynamics of a micro-particle in non-relativistic situations. It is also used to solve for the allowed energy levels of the particle.

1.6   Higher order PDEs

(1) Korteweg-de Vries equation (shallow water wave problem)

$$\phi_t = 6\phi\phi_x - \phi_{xxx}$$

It is a nonlinear PDE and tells of an interesting balance between nonlinearity and dispersion to cause appearance of soliton-like solutions.

(2) Fourth order diffusion equation

$$\phi_t = -\phi_{xxxx}$$

Fourth order PDEs appear in many places of physical problems such as in thin film models, surface diffusion on solids, etc. Perhaps the simplest example of a fourth order diffusion equation is the above form where the minus sign makes it a dissipative equation. Such a PDE is generated by the energy functional $H = \int_a^b \phi_x^2 dx$ on $L^2((a, b))$. Solving a forth or higher order PDE analytically is not always an easy task because of the difficulty in implementing the boundary conditions which are more in number as compared to the second order equations. Thus, one often has to take recourse to computational calculations.

From a class of solutions that a PDE may enjoy, imposition of boundary conditions or initial conditions or both picks out the one which is relevant to the problem at hand. A PDE when subjected to initial conditions constitutes an initial value problem while if boundary conditions are imposed, we have a boundary value problem. If both types of conditions are necessary then we speak of an initial-boundary value problem. One has to take care that adequate numbers of such conditions are prescribed otherwise we may not arrive at any solution at all or run into too many solutions. The solution we generally look for needs to have a unique character, depend continuously on the prescribed data (otherwise the solution will not be stable) and has to be well posed.

For the Poisson or Laplace's equation, time is not involved and so we have a boundary value problem at disposal. We normally take for the hypersurface a closed curve or surface (see Figure 1.2). The type of boundary value problem is distinguished by the nature of data that is prescribed. For instance, if ϕ is prescribed on the boundary of a closed curve or surface it is a Dirichlet's problem. On the other hand, if the normal derivative of ϕ is prescribed on the boundary of a closed curve or surface, we run into a Neumann problem. Sometimes a problem requires specification of the function ϕ on certain part of the hypersurface and the normal derivative of ϕ in the remaining part of the hypersurface. In such a case we have a mixed or Robin condition.

There is also an additional possibility when both ϕ and $\frac{\partial \phi}{\partial n}$ are specified on the hypersurface. We then have the Cauchy problem. In fact, for the wave equation which is an evolution equation, time is an open boundary and so one needs both ϕ and its derivative at an initial time. To these are to be supplemented by the values of ϕ at the boundary which is closed (see Figure 1.3).

For the heat conduction or diffusion equation time is an open boundary too. However, it is a first order equation and so the initial value of ϕ is sufficient. Of course since the spatial boundary is closed ϕ needs to be prescribed (see Figure 1.4).

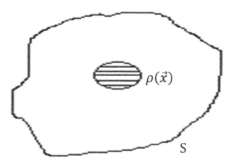

FIGURE 1.2: A closed curve with a boundary.

In general it is not possible to solve a Dirichlet problem for the wave equation nor a Cauchy problem for the Laplace or Poisson equation. This is a no-go situation. These will be illustrated with examples in Chapter 2.

1.7 Cauchy problem for second order linear PDEs

Let us adopt the following form of a second order linear PDE

$$A\phi_{xx} + 2B\phi_{xy} + C\phi_{yy} + D\phi_x + E\phi_y + F\phi = 0 \qquad (1.30)$$

where the coefficients A, B, C, D, E, F are functions of the two independent variables x and y possessing sufficiently many derivatives. Let us use the notations

$$p = \phi_x, \quad q = \phi_y, \quad r = \phi_{xx}, \quad s = \phi_{xy}, \quad t = \phi_{yy} \qquad (1.31)$$

This puts the above PDE in the following form

$$Ar + 2Bs + Ct = \Phi \qquad (1.32)$$

where Φ contains at most first order partial derivative terms.

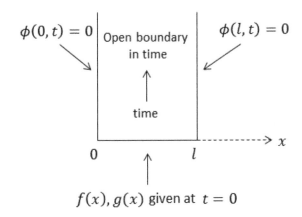

FIGURE 1.3: Open time boundary for the wave equation.

Suppose we are given a curve Γ along which both ϕ and its normal derivative $\frac{\partial \phi}{\partial n}$ are prescribed at some initial time. The Cauchy problem asks the question whether a solution of the PDE (1.30) exists that satisfies these initial conditions. Since knowing ϕ on Γ implies that the partial derivative with respect to the arc length $\frac{\partial \phi}{\partial s}$ is known, we can therefore claim[2] that along it ϕ as well as its partial derivatives ϕ_x and ϕ_y are also known.

The differentials which follow from (1.31) namely

$$dp = rdx + sdy, \quad dq = sdx + tdy \qquad (1.33)$$

also hold on Γ. From (1.32) and (1.33) we thus have three equations which, in principle, should determine r, s and t unless the coefficient determinant given by

$$\Delta = Ady^2 - 2Bdxdy + Cdy^2 \qquad (1.34)$$

vanishes.

The vanishing of Δ however always takes place if the derivative $\frac{dy}{dx}$ obeys

$$\frac{dy}{dx} = \frac{B \pm \sqrt{J}}{AC}, \quad J = B^2 - AC \qquad (1.35)$$

From (1.35) we can determine two directions for which $\Delta = 0$. These represent two families of curves which are called the characteristic curves for the second order system. Thus the solvability of the problem demands that Γ

---

[2]See the discussion in Chapter 2.

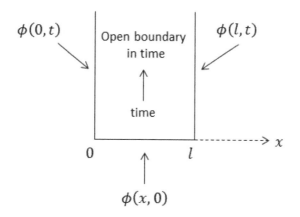

FIGURE 1.4: Open time boundary for the heat conduction equation.

shall nowhere be a tangent to the characteristic. Should this be so (i.e. Γ is a non-characteristic) we can undertake a Taylor series expansion as we did in the first order case and evaluate all the terms corresponding to various orders of derivatives of ϕ arising in it at some given point in an arbitrarily chosen neighbourhood in the xy-plane and thus uniquely determine the solution.

A second order linear PDE is classified by the sign of the quantity J. When $J < 0$, the PDE is of elliptic type and the characteristics are conjugate complex in character.

Real characteristics forming two distinct families of curves appear when $J > 0$. The PDE is then classified as of hyperbolic type.

Only one real family of a characteristic curve exists when $J = 0$ in which case the PDE is said to be of parabolic type.

In general, A, B and C are functions of x and y. So in the xy-plane there could be various regions where different types of characteristics occur for a given PDE.

The method of characteristics is a powerful tool for solving a PDE. Let us illustrate it by considering the following examples.

Example 1.11

Find the characteristics of the following PDE

$$\phi_{xx}(x, y) + xy\phi_{yy}(x, y) = 0$$

Hence transform the equation in terms of the characteristic variable.

This is a typical PDE in which it is of one type (hyperbolic) in a certain region of xy-plane and different type (elliptic) in another region of xy-plane. In fact, since here $A = 1$, $B = 0$ and $C = xy$, the quantity J is

$$J = -xy$$

signifying that the PDE is hyperbolic in the second and fourth quadrants of the xy-plane but elliptic in the first and third quadrant of the xy-plane.

The characteristic curves follow from the equation

$$(\frac{dy}{dx})^2 + xy = 0 \quad \rightarrow \quad \frac{dy}{dx} = \pm\sqrt{-xy}$$

We first focus on the hyperbolic case and take the second quadrant ($x < 0$, $y > 0$):

$$\frac{dy}{dx} = \sqrt{-x}\sqrt{y}$$

Simple integration furnishes the solution

$$y^{\frac{1}{2}} - \frac{1}{3}(-x)^{\frac{3}{2}} = \text{constant}$$

We now set

$$\xi = y^{\frac{1}{2}}, \quad \eta = \frac{1}{3}(-x)^{\frac{3}{2}}$$

which implies

$$\eta + \xi = \text{constant}, \quad \eta - \xi = \text{constant}$$

In terms of ξ and η the quantities ϕ_{xx} and ϕ_{yy} can be evaluated using the chain rule of differentiation:

$$\phi_{xx} = -\frac{x}{4}(\phi_{\eta\eta} + \frac{\phi_\eta}{3\eta}), \quad \phi_{yy} = \frac{1}{4y}(\phi_{\xi\xi} - \frac{\phi_\xi}{\xi})$$

Hence the given PDE becomes in the hyperbolic case

$$\phi_{\eta\eta} - \phi_{\xi\xi} + \frac{\phi_\eta}{3\eta} + \frac{\phi_\xi}{\xi} = 0$$

in terms of ξ and η variables. As we will learn in a later chapter this refers to the normal form of the equation in terms of the characteristic variables (ξ, η).

In a similar way, for the elliptic case we set

$$\xi = y^{\frac{1}{2}}, \quad \eta = \frac{1}{3}(x)^{\frac{3}{2}}$$

which implies

$$\eta + i\xi = \text{constant}, \quad \eta - i\xi = \text{constant}$$

Proceeding as in the hyperbolic case we get for the elliptic counterpart the corresponding normal form

$$\phi_{\eta\eta} + \phi_{\xi\xi} + \frac{\phi_\eta}{3\eta} - \frac{phi_\xi}{\xi} = 0$$

in terms of ξ and η variables.

Example 1.12

Solve the two-dimensional Laplace's equation

$$\phi_{xx}(x, y) + \phi_{yy}(x, y) = 0$$

within a square of length a subject to the boundary conditions

$$\phi(0, y) = 0, \quad \phi(a, y) = 0, \quad \phi(x, 0) = 0, \quad \phi(x, a) = \phi_0 \sin \frac{\pi x}{a}$$

With the conditions given on the boundary of the square, let us note that the problem is a well-posed one. There can be various methods to solve it but let us proceed with the method of characteristics.

Since the equation is of elliptic in nature the characteristics are conjugate complex with the derivative $\frac{dy}{dx}$ satisfying

$$\frac{dy}{dx} = 0$$

Indeed these provide the characteristic variables

$$\xi = x + iy = \lambda, \quad \eta = x - iy = \mu$$

where λ and μ are constants. In terms of ξ and η the Laplace's equation assumes the normal form

$$\frac{\partial^2 \phi}{\partial \xi \partial \eta} - 0$$

Its solution can be expressed in terms of two arbitrary functions χ and ψ i.e. $\phi(\xi, \eta) = \chi(\xi) + \psi(\eta)$. Projecting in terms of the variables x and y we can express ϕ as

$$\phi(x, y) = \chi(x + iy) + \psi(x - iy)$$

To be justified as a solution of the Laplace's equation we need to fix χ and ψ properly. Keeping in mind the boundary conditions, the following periodic choice is natural

$$\phi(x,y) = ik(\cos[\frac{\pi}{a}(x+iy)] - \cos[\frac{\pi}{a}(x-iy)])$$

where k is an overall constant. We keep the real part

$$\phi(x,y) = k\sin(\frac{\pi x}{a})\sinh(\frac{\pi y}{a})$$

All except the boundary conditions $\phi(x,a) = \phi_0 \sin\frac{\pi x}{a}$ are fulfilled. We come to terms with this one by fixing $k = \frac{\phi_0}{\sinh\pi}$. It gives the required solution

$$\phi(x,y) = \frac{\phi_0}{\sinh\pi}\sin(\frac{\pi x}{a})\sinh(\frac{\pi y}{a})$$

Example 1.13

Find the characteristics of the parabolic equation

$$\phi_t(x,t) = \alpha^2\phi_{xx}(x,t)$$

where α is a constant. Transform the equation in terms of the characteristic variable.

Here $A = \alpha^2$, $B = 0 = C$ implying $J = 0$. Therefore the characteristic equation is

$$\frac{dt}{dx} = 0$$

We thus see that the $t = c$ lines represent the characteristics. We can transform the variables according to

$$\xi = t, \quad \eta = x$$

and notice that the equation is already in normal form.

1.8 Hamilton-Jacobi equation

In this section we focus on the occurrence of a typical linear first order PDE which is of immense utility in problems of classical mechanics. It is called the Hamilton-Jacobi equation. As is well known the analytical structure of classical dynamics describing the motion of a particle or a system of particles rests on two important but almost equivalent principles namely, those of Lagrangian and Hamiltonian mechanics. While the Lagrangian for system of n degrees of freedom is a function of the generalized coordinates $q_i, i = 1, 2, ..., n$ and generalized velocities $\dot{q}_i, i = 1, 2, ..., n$ and appears as the difference of kinetic and potential energies of the particles, the Hamiltonian, which can be derived from the Lagrangian by effecting a Legendre transformation, is regarded as a function of the generalized coordinates $q_i, i = 1, 2, ..., n$ and generalized momenta $p_i, i = 1, 2, ..., n$. It appears, in a conservative system, as the sum of kinetic and potential energies. The underlying equations of motion are given by a set of n first order partial differential equations each for the time derivative of the generalized coordinates and momenta. The Hamiltonian procedure is carried out in a phase space which is $2n$-dimensional being made up of positions and the accompanying canonical momenta. The Hamiltonian equations give the clue as to how a system evolves in phase space. Both Lagrangian and Hamiltonian methods address the laws of mechanics without any preference to the selection of any particular coordinate system.

Use of Hamilton's equations does not always offer any simplification to the understanding of the concerned dynamics. But the almost symmetrical appearance of the coordinates and momenta in them facilitates development of formal theories such as the canonical transformations and Hamilton-Jacobi equation. We will discuss, in the following, how canonical transformations are constructed that offer considerable simplifications in writing down the equations of motion. The task is to transform the Hamiltonian to a new form through the introduction of a new set of coordinates and momenta such that with respect to them the form of canonical equations is preserved. The advantage of employing canonical transformation lies in the fact that it is often possible to adopt new sets of conjugate variables through which the basic equations get more simplified thus generating solutions which may otherwise be very complicated to determine. The special case when all the coordinates are cyclic yields a single partial differential equation of first order which is called the Hamilton-Jacobi equation.

1.9   Canonical transformation

For simplicity we will be frequently restricting to a system of one degree of freedom described by a single pair of canonical variables (q, p). The phase space is then of dimension two. The Hamiltonian equations are given by

$$\dot{q} = \frac{\partial H}{\partial p}, \quad \dot{p} = -\frac{\partial H}{\partial q} \tag{1.36}$$

We ask the question whether a transformation from (q, p) to a new set of variables (Q, P) is feasible in a way that the above form of the Hamiltonian equations is preserved

$$\dot{Q} = \frac{\partial K}{\partial P}, \quad \dot{P} = -\frac{\partial K}{\partial Q} \tag{1.37}$$

We emphasize that our intention in looking for new variables (Q, P) is to inquire if these could also serve in a new but simplified means of describing the system but at the same time not disturb the essential physics content. In (1.36), K is a transformed version of H

$$H(q(Q, P), p(Q, P)) \to K(Q, P) \equiv K \tag{1.38}$$

defined in terms of the variable Q and P. It is obvious that the equations (1.36) may not always hold except for some special situations. Such restricted transformations for which these equations hold are called canonical transformations implying that the new variables (Q, P) too form a canonical set.

Noting that the Poisson bracket of any two functions F and G is defined to be

$$\{F, G\}_{(q,p)} = \frac{\partial F}{\partial q} \frac{\partial G}{\partial p} - \frac{\partial F}{\partial p} \frac{\partial G}{\partial q} \tag{1.39}$$

where (q, p) is the set of canonical variables, Hamiltonian equations for $Q = Q(q, p), P = P(q, p)$ read

$$\dot{Q} = \{Q, H\}_{(q,p)} = \frac{\partial Q}{\partial q} \frac{\partial H}{\partial p} - \frac{\partial Q}{\partial p} \frac{\partial H}{\partial q}$$

$$\dot{P} = \{P, H\}_{(q,p)} = \frac{\partial P}{\partial q} \frac{\partial H}{\partial p} - \frac{\partial P}{\partial p} \frac{\partial H}{\partial q} \tag{1.40}$$

Interpreting $H(q, p)$ as $K(Q, P)$ in terms of the new variables (Q, P) and employing the chain rule of partial derivatives gives

$$\frac{\partial H}{\partial q} = \frac{\partial K}{\partial Q} \frac{\partial Q}{\partial q} + \frac{\partial K}{\partial P} \frac{\partial P}{\partial q}$$

$$\frac{\partial H}{\partial p} = \frac{\partial K}{\partial Q} \frac{\partial Q}{\partial p} + \frac{\partial K}{\partial P} \frac{\partial P}{\partial p} \tag{1.41}$$

As a result we are led to the following expressions for $\dot{Q}$ and $\dot{P}$

$$
\begin{aligned}
\dot{Q} &= \frac{\partial Q}{\partial q}\left(\frac{\partial K}{\partial Q}\frac{\partial Q}{\partial p}+\frac{\partial K}{\partial P}\frac{\partial P}{\partial p}\right)-\frac{\partial Q}{\partial p}\left(\frac{\partial K}{\partial Q}\frac{\partial Q}{\partial q}+\frac{\partial K}{\partial P}\frac{\partial P}{\partial q}\right) \\
&= \frac{\partial K}{\partial P}\left(\frac{\partial Q}{\partial q}\frac{\partial P}{\partial p}-\frac{\partial Q}{\partial p}\frac{\partial P}{\partial q}\right) \\
&= \frac{\partial K}{\partial P}\{Q,P\}_{(q,p)} \\
\dot{P} &= -\frac{\partial K}{\partial Q}\{Q,P\}_{(q,p)}
\end{aligned}
\tag{1.42}
$$

These equations coincide with (1.37) provided we set $\{Q,P\}_{(q,p)}$ equal to unity:

$$
\{Q,P\}_{(q,p)} = 1 \tag{1.43}
$$

Hamiltonian forms of canonical equations are then also valid for the new pair of variables (Q,P).

For a system of n degrees of freedom the condition for the canonical transformation is given by the Poisson bracket conditions

$$
\{Q_i,Q_j\}_{(q,p)} = 0, \quad \{P_i,P_j\}_{(q,p)} = 0, \quad \{Q_i,P_j\}_{(q,p)} = \delta_{ij} \tag{1.44}
$$

The invariance of Poisson bracket relations is a fundamental feature of canonical transformations. In fact the above properties of the Poisson brackets serve as the necessary and sufficient conditions for a transformation to be canonical. Note that out of $4n$ variables, (q_i,p_i) and (Q_i,P_i), $i = 1,2,...,n$, only $2n$ of these are independent.

It needs to be pointed out that the Poisson bracket $\{Q,P\}_{(q,p)}$ is the same as the Jacobian determinant :

$$
\begin{aligned}
\{Q,P\}_{(q,p)} &= \frac{\partial Q}{\partial q}\frac{\partial P}{\partial p}-\frac{\partial Q}{\partial p}\frac{\partial P}{\partial q} \\
&= \frac{\partial(Q,P)}{\partial(q,p)}
\end{aligned}
\tag{1.45}
$$

Conversely we also have

$$
\{q,p\}_{(Q,P)} = \frac{\partial(q,p)}{\partial(Q,P)} = \left[\frac{\partial(Q,P)}{\partial(q,p)}\right]^{-1} = \left[\{Q,P\}_{(q,p)}\right]^{-1} \tag{1.46}
$$

Consider a region D in the phase space plane (q,p) that is bounded by a closed curve A. Using (1.45) we can express

$$\int_D dQdP = \int_D \frac{\partial(Q,P)}{\partial(q,p)}dqdp = \int_D [Q,P]_{q,p}dqdp = \int_D dqdp \qquad (1.47)$$

where we have exploited (1.43). In terms of closed line integrals we then have by Stokes's theorem

$$\oint_A pdq = \oint_A PdQ \qquad (1.48)$$

In general, for a system with n generalized coordinates q_i together with their associated momenta p_i, the volume element transforms as

$$\prod_{i=1}^{n} dQ_i dP_i = \Delta \prod_{i=1}^{n} dq_i dp_i = \prod_{i=1}^{n} dq_i dp_i \qquad (1.49)$$

where Δ stands for the determinant of the Jacobian matrix

$$\Delta = \det[\frac{\partial(Q_1,Q_2,...,Q_n;P_1,P_2,...,P_n)}{\partial(q_1,q_2,...,q_n;p_1,p_2,...,p_n)}] \qquad (1.50)$$

and we employed Liouville's theorem which points to the preservation of volume in the phase space. Thus phase space volume remains invariant under canonical transformations.

Example 1.14

Consider the transformation $(q,p) \to (Q,P)$ given by

$$q = \sqrt{\frac{2P}{m\omega}} \sin Q, \quad p = \sqrt{2m\omega P} \cos Q$$

Inverting

$$Q = \tan^{-1}\left(m\omega\frac{q}{p}\right), \quad P = \frac{1}{2\omega}\left(\frac{p^2}{m} + m\omega^2 q^2\right)$$

$$\Rightarrow \frac{\partial Q}{\partial q} = \frac{m\omega p}{p^2 + m^2\omega^2 q^2}, \quad \frac{\partial Q}{\partial p} = -\frac{m\omega q}{p^2 + m^2\omega^2 q^2}$$

As a result

$$\begin{aligned}
\{Q,P\}_{(q,p)} &= \frac{\partial Q}{\partial q}\frac{\partial P}{\partial p} - \frac{\partial Q}{\partial p}\frac{\partial P}{\partial q} \\
&= \frac{1}{p^2 + m^2\omega^2 q^2}\left(m\omega p\frac{\partial P}{\partial p} + m\omega q\frac{\partial P}{\partial q}\right)
\end{aligned}$$

Since $\frac{\partial P}{\partial p} = \frac{p}{m\omega}$, $\frac{\partial P}{\partial q} = m\omega q$ it follows that $\{Q, P\}_{(q,p)} = 1$. Hence the transformation is canonical.

Example 1.15: Harmonic oscillator problem

Let us now concentrate on the specific case of the harmonic oscillator described by the Hamiltonian $H = \frac{1}{2m}p^2 + \frac{1}{2}m\omega^2 q^2$. It is clear from the form of P in the previous example that if employed the Hamiltonian takes a very simple form $H \to K = P\omega$, where K is the transformed Hamiltonian. The accompanying Hamilton's equations for the new variables Q and P follow easily

$$\dot{Q} = \frac{\partial K}{\partial P} = \omega$$

$$\dot{P} = -\frac{\partial K}{\partial Q} = 0$$

Solving for Q and P we find $Q = \omega t + t_0$, $\quad P = b$ where t_0 and b are constants of integration. Switching to the original variables we get for q and p

$$q = \sqrt{\frac{2b}{m\omega}} \sin\left(\omega t + t_0\right)$$
$$p = \sqrt{2m\omega b} \cos\left(\omega t + t_0\right)$$

which conform to their standard forms. This example serves to illustrate the advantage of employing canonical variables judiciously that allow the basic equations to get simplified and so obtaining the solutions becomes an easy task.

1.10   Concept of generating function

That the quantity $(pdq - PdQ)$ is an exact differential can be established by noting

$$
\begin{aligned}
pdq - PdQ &= pdq - P\left(\frac{\partial Q}{\partial q}dq + \frac{\partial Q}{\partial p}dp\right) \\
&= \left(p - P\frac{\partial Q}{\partial q}\right)dq - P\frac{\partial Q}{\partial p}dp
\end{aligned}
\tag{1.51}
$$

from which the condition for an exact differential namely,

$$\frac{\partial}{\partial p}\left(p - P\frac{\partial Q}{\partial q}\right) = \frac{\partial}{\partial q}\left(-P\frac{\partial Q}{\partial p}\right) \tag{1.52}$$

holds because of $\{Q, P\}_{(q,p)} = 1$.

Setting $pdq - PdQ = dG_1$, we call G_1 to be the generating function of the transformation $(q, p) \rightarrow (Q, P)$. Such a class of generating functions is referred to as the Type I generating function. G_1 being a function of q and Q [i.e. $G_1 = G_1(q, Q)$] gives

$$dG_1 = \frac{\partial G_1}{\partial q}dq + \frac{\partial G_1}{\partial Q}dQ \tag{1.53}$$

Matching with the left hand side of (1.51)

$$p = \frac{\partial G_1}{\partial q}, \quad P = -\frac{\partial G_1}{\partial Q} \tag{1.54}$$

Consider the case of Example 1.14. We first of all check whether $pdq - PdQ$ is a perfect differential for this problem. For the generating function G_1, the independent variables are q and Q. Expressing p and P in terms of these namely, $p = m\omega q \cot Q$, $P = \frac{1}{2}m\omega q^2 \mathrm{cosec}^2 Q$ we can express

$$pdq - PdQ = d\left(\frac{1}{2}m\omega q^2 \cot\theta\right) \tag{1.55}$$

which is indeed an exact differential. Hence the generating function G_1 for this problem is $G_1(q, Q) = \frac{1}{2}m\omega q^2 \cot\theta$.

There can be other types of generating functions depending on what combinations of old and new canonical variables we choose. For instance a Type II generating function G_2 is a function of q and P and the counterpart of (1.54) reads

$$p = \frac{\partial G_2}{\partial q}, \quad Q = \frac{\partial G_2}{\partial P} \tag{1.56}$$

For the Type III generating function G_3 which is a function of p and Q the underlying relations are

$$q = -\frac{\partial G_3}{\partial p}, \quad P = -\frac{\partial G_3}{\partial Q} \tag{1.57}$$

Finally, a Type IV generating function G_4 depends on the pair (p, P) with the complementary variables q and Q satisfying

$$q = -\frac{\partial G_4}{\partial p}, \quad Q = -\frac{\partial G_4}{\partial P} \tag{1.58}$$

The transformations given in (1.54), (1.56), (1.57) and (1.58) hold for the time-independent canonical transformations. These are *restricted* canonical transformations. In the time-dependent case, which we discuss in the next section, the generating functions possess additionally a function of the variable t.

1.11 Types of time-dependent canonical transformations

Consider a transformation $(q, p, t) \to (Q, P, t)$ for a general time-dependent situation. Since adding a total differential does not change the essential dynamics of the system, we define a time-dependent canonical transformation $(q, p, t) \to (Q, P, t)$ according to the following criterion:

$$\sum_{i=1}^{n} p_i dq_i - H dt = \sum_{i=1}^{n} P_i dQ_i - K dt + dG_1 \tag{1.59}$$

with $G_1 = G_1(q, Q, t)$ to be the time-dependent Type I generating function of the transformation. The time-independent generating function was considered earlier from the exact differentiability of the quantity $(pdq - PdQ)$.

We now turn to four kinds of canonical transformations as induced by the corresponding time-dependent generating functions.

1.11.1 Type I Canonical transformation

A Type I canonical transformation treats the old and new coordinates q_i and Q_i as independent variables. For $G_1(q_i, Q_i, t)$ we have

$$dG_1 = \sum_{i=1}^{n} \left(\frac{\partial G_1}{\partial q_i} dq_i + \frac{\partial G_1}{\partial Q_i} dQ_i \right) + \frac{\partial G_1}{\partial t} dt \tag{1.60}$$

Putting this form in (1.59) we find on comparing the differentials

$$p_i = \frac{\partial G_1(q_i, Q_i, t)}{\partial q_i}, \quad P_i = -\frac{\partial G_1(q_i, Q_i, t)}{\partial Q_i} \tag{1.61}$$

along with

$$K = H + \frac{\partial G_1}{\partial t} \tag{1.62}$$

The first of (1.61) gives Q_i in terms of q_i and p_i which when substituted in the second equation determines P_i. We of course assume that the Jacobian of the transformations det $|\frac{\partial^2 G_1}{\partial q_i \partial Q_j}| \neq 0$). K is the new Hamiltonian.

It is worthwhile to note that the Type I canonical transformation induces the exchange transformation. For instance if we choose $G_1(q, Q, t) = q_i Q_i$ then p_i and P_i turn out to be Q_i and $-q_i$ respectively while $K = H$ reflects that the coordinates and momenta are exchanged.

1.11.2   Type II Canonical transformation

For a Type II canonical transformation the independent variables are q_i and P_i which stand respectively for the old coordinates and the new momenta. In this case we express

$$\sum_{i=1}^{n} P_i dQ_i = d\left(\sum_{i=1}^{n} P_i Q_i\right) - \sum_{i=1}^{n} Q_i dP_i \tag{1.63}$$

which results in

$$\sum_{i=1}^{n} p_i dq_i + \sum_{i=1}^{n} Q_i dP_i - H dt = d\left(\sum_{i=1}^{n} P_i Q_i\right) - K dt + dG_1 \tag{1.64}$$

The accompanying generating function G_2 can therefore be defined by

$$G_2 = G_1 + \sum_{i=1}^{n} P_i Q_i \tag{1.65}$$

Viewing G_2 as a function of q_i, P_i and t we write

$$dG_2(q_i, P_i, t) = \sum_{i=1}^{n} \left(\frac{\partial G_2}{\partial q_i} dq_i + \frac{\partial G_2}{\partial P_i} dP_i\right) + \frac{\partial G_2}{\partial t} dt \tag{1.66}$$

which implies from (1.59) and (1.65)

$$p_i = \frac{\partial G_2(q_i, P_i, t)}{\partial q_i}, \quad Q_i = \frac{\partial G_2(q_i, P_i, t)}{\partial P_i} \tag{1.67}$$

along with

$$K = H + \frac{\partial G_2}{\partial t} \tag{1.68}$$

The first of (1.67) gives P_i in terms of q_i and p_i which when substituted in the second equation provides for Q_i. The Jacobian of the transformation is assumed to be nonvanishing: $\det |\frac{\partial^2 G_2}{\partial q_i \partial P_j}| \neq 0$. K as given by (1.68) is the transformed Hamiltonian.

The Type II canonical transformation has in its embedding both the identity as well as the point transformations. In fact, corresponding to $G_2(q, P, t) = q_i P_i$ we see that $p_i = P_i$, $Q_i = q_i$ and $K = H$ showing for the identity transformation while for $G_2(q, P, t) = \phi(q_i, t)P_i$ we have $Q_i = \phi(q_i, t)$ implying that the new coordinates are functions of old coordinates.

1.11.3 Type III Canonical transformation

In Type III canonical transformation the underlying generating function G_3 is a function of the independent variables p_i, Q_i which are respectively the old momenta and new coordinates. G_3 is defined by

$$G_3(p_i, Q_i, t) = G_1 - \sum_{i=1}^{n} q_i p_i \tag{1.69}$$

which leads to

$$q_i = -\frac{\partial G_3}{\partial p_i}, \quad P_i = -\frac{\partial G_3}{\partial Q_i}, \quad K = H + \frac{\partial G_3}{\partial t} \tag{1.70}$$

Here the Jacobian of the transformation is assumed to be nonvanishing: $\det |\frac{\partial^2 G_3}{\partial p_i \partial Q_j}| \neq 0$.

1.11.4 Type IV Canonical transformation

In the Type IV canonical transformation the generating function is a function of old and new momenta and given by

$$G_4(p_i, P_i, t) = G_1 + \sum_{i=1}^{n} Q_i P_i - \sum_{i=1}^{n} q_i p_i \tag{1.71}$$

in which p_i and P_i are treated as independent variables. We then have

$$q_i = -\frac{\partial G_4}{\partial p_i}, \quad Q_i = \frac{\partial G_4}{\partial P_i}, \quad K = H + \frac{\partial G_4}{\partial t} \tag{1.72}$$

Here the Jacobian of the transformation is assumed to be nonvanishing: $\det |\frac{\partial^2 G_4}{\partial p_i \partial P_j}| \neq 0$.

Example 1.17

Consider the transformation

$$Q = -p, \quad P = q + \lambda p^2$$

where λ is a constant. By the Poisson bracket test namely $[Q, P]_{(q,p)} = 1$ we conclude that the transformation is a canonical transformation.

The Type I generating function is determined by showing $pdq - PdQ$ to be a perfect differential:

$$
\begin{aligned}
pdq - PdQ &= (-Q)dq - (q + \lambda Q^2)dQ \\
&= d(-qQ - \frac{1}{3}\lambda Q^3)
\end{aligned}
$$

Thus

$$G_1(q, Q) = -qQ - \frac{1}{3}\lambda Q^3$$

On the other hand the Type II generating function is obtained by treating q and P as independent variables. From (1.65) we have

$$
\begin{aligned}
G_2(q, p) &= G_1 + PQ \\
&= -qQ - \frac{1}{3}\lambda Q^3 + (q + \lambda Q^2)Q \\
&= \frac{2}{3}\lambda Q^3 \\
&= \frac{2}{3}\lambda \left(\frac{P - q}{\lambda}\right)^{\frac{3}{2}}
\end{aligned}
$$

Check that

$$
\begin{aligned}
\left(\frac{\partial G_2}{\partial q}\right)_P &= -\frac{1}{\sqrt{\lambda}}(P - q)^{\frac{1}{2}} = p \\
\left(\frac{\partial G_2}{\partial P}\right)_q &= \frac{1}{\sqrt{\lambda}}(P - q)^{\frac{1}{2}} = Q
\end{aligned}
$$

which are as required.

Example 1.18

The transformations

$$
\begin{aligned}
Q &= q\cos\theta - \frac{p}{m\omega}\sin\theta \\
P &= m\omega q\sin\theta + p\cos\theta
\end{aligned}
$$

are easily seen to be canonical due to $\{Q, P\}_{(q,p)} = 1$. We also find

$$
\begin{aligned}
p &= m\omega\left(q\cot\theta - \frac{Q}{\sin\theta}\right) \\
P &= m\omega\left(\frac{q}{\sin\theta} - Q\cot\theta\right)
\end{aligned}
$$

Hence the quantity $pdq - PdQ$ can be expressed as

$$pdq - PdQ = d\left[\frac{1}{2}m\omega(q^2 + Q^2)\cot\theta - m\omega qQ\text{cosec}\theta\right]$$

The Type I generating function $G_1(q, Q)$ is identified as

$$G_1(q, Q) = \frac{1}{2}m\omega(q^2 + Q^2)\cot\theta - m\omega qQ\text{cosec}\theta$$

On other other hand, the Type II generating function can be obtained from $G_2 = G_1 + PQ$. We get

$$G_2(q, Q) = \frac{1}{2}m\omega(q^2 - Q^2)\cot\theta$$

To assign the right variable dependence on G_2 namely q and P we note that

$$Q = \frac{q}{\cos\theta} - \frac{P}{m\omega}\tan\theta$$

by eliminating p from the given transformations. Substituting for Q, G_2 turns out to be

$$G_2(q, P) = \frac{qP}{\cos\theta} - \frac{1}{2}m\omega\left(q^2 + \frac{P^2}{m^2\omega^2}\right)\tan\theta$$

We can verify that

$$\left(\frac{\partial G_2}{\partial q}\right)_P = \frac{P}{\cos\theta} - m\omega q\tan\theta = p$$

$$\left(\frac{\partial G_2}{\partial P}\right)_q = \frac{q}{\cos\theta} - \frac{P}{m\omega}\tan\theta = Q$$

Example 1.19

We consider the harmonic oscillator Hamiltonian H which is invariant under the set of canonical transformations considered in the previous example:

$$H(q, p) = \frac{1}{2m}p^2 + \frac{1}{2}m\omega^2 q^2$$

$$\rightarrow \frac{1}{2m}P^2 + \frac{1}{2}m\omega^2 Q^2 = H(Q, P)$$

The generating function $G_2(q, P)$ which was found to be

$$G_2(q, P) = \frac{qP}{\cos\theta} - \frac{1}{2}m\omega\left(q^2 + \frac{P^2}{m^2\omega^2}\right)\tan\theta$$

gives for a partial derivative with respect to time

$$\left(\frac{\partial G_2}{\partial t}\right)_{q,P} = \left[qP\sin\theta - \frac{1}{2}m\omega\left(q^2 + \frac{p^2}{m^2\omega^2}\right)\right]\sec^2\theta\dot\theta$$

$$= -\left(\frac{P^2}{2m\omega} + \frac{1}{2}m\omega Q^2\right)\dot\theta$$

$$= -H(Q,P)\frac{\dot\theta}{\omega}$$

Therefore the transformed Hamiltonian reads

$$K(Q,P,t) = \left(1 - \frac{\dot\theta}{\omega}\right)H(Q,P)$$

which vanishes for $\dot\theta = \omega$ i.e. $\theta = \omega t$. As a consequence $\dot Q = \frac{\partial K}{\partial P} = 0$ and $\dot P = -\frac{\partial K}{\partial Q} = 0$. Hence Q and P are constants which we write as

$$Q = Q_0, \quad P = P_0$$

where Q_0 and P_0 stand for the initial values of q and p respectively.
Reverting to the old coordinates we get

$$q(t) = Q_0\cos\omega t + \frac{P_0}{m\omega}\sin\omega t$$
$$p(t) = -m\omega Q_0\sin\omega t + P_0\cos\omega t$$

which give the time evolution for q and p.

1.12   Derivation of Hamilton-Jacobi equation

The idea of deriving Hamilton-Jacobi equation rests on applying a canonical transformation that maps one basis of known canonical variables to a new set of coordinates and momenta such that the latter are cyclic with respect to the transformed Hamiltonian. What does it mean?

Consider the specific case of a Type II generating function $G_2(q_i, P_i, t), i = 1, 2, ..., n$ which facilitates the transformations according to (1.67) where the time derivatives of Q_i and P_i stand as

$$\dot Q_i = \frac{\partial K}{\partial P_i}, \quad \dot P_i = -\frac{\partial K}{\partial Q_i}, \quad i = 1, 2, ..., n \tag{1.73}$$

as defined by (1.37). Now if Q_i and P_i are cyclic coordinates in the transformed Hamiltonian K, then it follows that these have to be constants in time. Such a validity is assured if we assume, without loss of generality, K to be zero. It then gives from (1.68) the equation

$$H(q_1, q_2, ..., q_n; p_1, p_2, ..., p_n; t) + \frac{\partial G_2}{\partial t} = 0 \qquad (1.74)$$

A few remarks are in order:

From (1.66) if we take the time-derivative of the generating function $G_2(q_i, p_i, t)$ then in view of the above equation it follows that we can write

$$\frac{dG_2}{dt} = \sum_{i=1}^{n} \frac{\partial G_2}{\partial q_i} \dot{q}_i = \sum_{i=1}^{n} p_i \dot{q}_i - H(q_i, p_i, t) = L \qquad (1.75)$$

where we employed (1.67) and noted that P_i's are cyclic coordinates in K and hence treated as constants. Thus integrating between times say, t_1 and t_2, the quantity $S \equiv G_2$ defined by the action integral $S = \int_{t_1}^{t_2} L dt$ follows from (1.75). We therefore arrive at the form

$$p_i = \frac{\partial S}{\partial q_i} : \quad H(q_1, q_2, ..., q_n; \frac{\partial S}{\partial q_1}, \frac{\partial S}{\partial q_2}, ..., \frac{\partial S}{\partial q_n}; t) + \frac{\partial S}{\partial t} = 0 \qquad (1.76)$$

Equation (1.76) is called the time-dependent Hamilton-Jacobi equation.[3] More compactly it is also expressed as

$$H(q_i, \frac{\partial S}{\partial q_i}, t) + \frac{\partial S}{\partial t} = 0 \qquad (1.77)$$

A function S which is a solution of the above equation is often referred to as the Hamilton Principal function.

Equation (1.77) is first order differential and involves the n coordinates q_i's and t. It is not linear in $\frac{\partial S}{\partial q_i}$ because the partial derivatives of S appear in higher degree than the first. Associated with the $(n+1)$ variables (n q_i's and t) we expect $(n+1)$ constants of motion namely $\alpha_1, \alpha_2, ..., \alpha_n, \alpha_{n+1}$. However we notice one curious thing in (1.77) is that the dependent variable S itself does not appear in it: only its partial derivatives do. So one of the constants has no bearing on the solution *i.e.* the solution has an additive constant. Disregarding such an irrelevant additive constant we write for the complete integral of (1.77) the form

$$S = S(q_1, q_2, ..., q_n; \alpha_1, \alpha_2, ..., \alpha_n; t) \qquad (1.78)$$

---

[3]There could be other variants of Hamilton-Jacobi equation resulting Type I, Type III and Type IV generating functions.

where it is ensured that none of the constant α's is additive to the solution. Actually we identify these constants with the constant momenta P_i, $i = 1.2..., n$, which, according to our choice, were assumed to be cyclic.

The cyclic coordinates Q_i's provide another set of constants $\beta'_i s, i = 1, 2, ..., n$ and read from the second equation of (1.67)

$$Q_i = \frac{\partial S}{\partial \alpha_i} = \beta_i, \quad i = 1, 2, ..., n \qquad (1.79)$$

We clarify the above issues by considering two examples: one for the free particle and the other that of the harmonic oscillator.

Example 1.20: Free particle case

The Hamilton-Jacobi equation for the free particle problem is obviously

$$\frac{1}{2m}\left(\frac{\partial S}{\partial q}\right)^2 + \frac{\partial S}{\partial t} = 0$$

Its complete integral can be easily ascertained by inspection which reads

$$S(q, \alpha, t) = \sqrt{2m\alpha}q - \alpha t$$

where α is a non-additive constant.

From (1.79)

$$\beta = \frac{\partial S}{\partial \alpha} = k \quad \text{(say)}$$

i.e. $\quad k = \sqrt{\frac{m}{2\alpha}}q - t \quad \Rightarrow \quad q = \sqrt{\frac{2E}{m}}(t + k) \quad \text{and} \quad p = \frac{\partial S}{\partial q} = \sqrt{2m\alpha}.$

which conform to their expected forms.

It is to be noted that Hamilton-Jacobi equation being a partial differential equation can admit of multiple solutions. For instance, in the above case, there is also a legitimate solution given by

$$S(q, \alpha, t) = \frac{m(q - \alpha)^2}{2t}$$

where α is a non-additive constant. This implies

$$\beta = \frac{\partial S}{\partial \alpha} = -\frac{m}{t}(q - \alpha)$$

giving the time-dependence: $q(t) = \alpha - \frac{\beta}{m}t$ where α can be identified as the initial value of q while β is the negative of the momentum: $\beta = -p$.

Example 1.22

For the harmonic oscillator problem the Hamilton-Jacobi equation has the form

$$H = \frac{1}{2m} \left(p^2 + m^2 \omega^2 q^2 \right)$$

The relation $p = \frac{\partial S}{\partial q}$ gives

$$\frac{1}{2m} \left[\left(\frac{\partial S}{\partial q} \right)^2 + m^2 \omega^2 q^2 \right] + \frac{\partial S}{\partial t} = 0$$

To solve for the above equation we try separation of variables for S:

$$S(q : \alpha; t) = W(q : \alpha) - \alpha t$$

where α is a non-additive constant. The function $W(q : \alpha)$ is the time-independent part called the Hamilton's characteristic function. We get

$$\frac{1}{2m} \left[\left(\frac{\partial W}{\partial q} \right)^2 + m^2 \omega^2 q^2 \right] = \alpha$$

and gives the integral

$$W = \sqrt{2m\alpha} \int dq \sqrt{1 - \frac{m\omega^2 q^2}{2\alpha}}$$

It implies for β

$$\beta = \frac{\partial S}{\partial \alpha} = \sqrt{\frac{m}{2\alpha}} \int \frac{dq}{\sqrt{1 - \frac{m\omega^2 q^2}{2\alpha}}} - t$$

which easily integrates to

$$t + \beta = \frac{1}{\omega} \sin^{-1} q \sqrt{\frac{m\omega^2}{2\alpha}}$$

Hence the solutions for q and p are

$$q = \sqrt{\frac{2\alpha}{m\omega^2}} \sin \omega(t + \beta)$$

$$p = \frac{\partial S}{\partial q} = \frac{\partial W}{\partial q} = \sqrt{2m\alpha - m\omega^2 q^2} = \sqrt{2m\alpha} \cos \omega(t + \beta)$$

which are agreement with their well known forms.

1.13   Summary

In this chapter we began by familiarizing the readers with some of the background formalities of a PDE with a view to fixing the notations and explaining them by accompanying examples. We showed how a PDE can be generated by some elementary means namely, elimination of constants or elimination of an arbitrary function. We pursued in some detail the linear first order equation, in particular we dealt with the quasi-linear case. We then gave some examples of second and higher order PDE and discussed the Cauchy problem at length. The method of characteristics was illustrated by means of examples. Finally, from an application point of view, we discussed the canonical transformations along with the Hamilton-Jacobi equation which is a linear first order PDE and of immense importance in classical mechanics.

Exercises

1. Find the general solution of the PDE

$$x\phi\phi_x + y\phi\phi_y + (x^2 + y^2) = 0$$

2. Find the general solution of the linear equation

$$(y - z)\frac{\partial\phi}{\partial x} + (z - x)\frac{\partial\phi}{\partial y} + (x - y)\frac{\partial\phi}{\partial z} = 0$$

3. Find the general solution of the linear equation

$$2x\phi_x + 2y\phi\phi_y = \phi^2 - x^2 - y^2$$

Also determine the integral surface which passes through the circle

$$x + y + \phi = 0, \quad x^2 + y^2 + z^2 = a^2$$

where a is constant.

4. Solve the following PDE

$$(\phi + y + z)\phi_x + (\phi + z + x)\phi_y + (\phi + x + y)\phi_z = x + y + z$$

by employing Lagrange's method.

5. Consider the quasi-linear equation

$$x(y^2 + z)\phi_x - y(x^2 + z)\phi_y = (x^2 y^2)\phi$$

Determine the integral surface which passes through the line $x + y = 0, z = 1$.

6. Show that integral surface of the PDE

$$(x - y)y^2\phi_x + (y - x)x^2\phi_y = (x^2 + y^2)\phi$$

which contains the straight line $x + y = 0, \phi = 1$ is given by the form $f(x^2 + y\phi, x^2 + y^2) = 0$, where f is an arbitrary function.

7. Show that the integral surface of the PDE

$$x(y^2 + \phi)\phi_x - y(x^2 + \phi) = (x^2 - y^2)\phi$$

which passes through the curve $x\phi = a^3, y = 0$ is given by the form $\phi^3(x^3 + y^3)^2 = a^9(x - y)^3$, where a is a constant.

8. Show that a local solution of the partial differential equation $\phi_t + a\phi_x = \phi^2$ subject to $\phi(x, 0) = \cos(x)$, where a is a constant can be written as

$$\phi(x, t) = \frac{\cos(x - at)}{1 - t\cos(x - at)}$$

9. Find the characteristic curves and the general solution of the linear equation

$$\phi_x + e^x\phi_y + e^z\phi_z = (2x + e^x)e^\phi$$

10. Find the characteristics of the following PDE

$$\phi_{xx} + y\phi_{yy} = 0$$

If the characteristic variables are ξ and η for the hyperbolic case show that the transformed equation is

$$\phi_{\xi\eta} + \frac{1}{\eta - \xi}(\phi_\eta - \phi_\xi) = 0$$

On the other hand, if the characteristic variables are σ and β for the elliptic case show that the transformed equation is

$$\phi_{\sigma\sigma} + \phi_{\beta\beta} - \frac{1}{\beta}\phi_\beta = 0$$

Explain the significance of the singularity at $\eta = \xi$ for the hyperbolic case and $\beta = 0$ for the elliptic case.

Chapter 2

Basic properties of second order linear PDEs

2.1   Preliminaries

Given a domain Ω in the n-dimensional Euclidean space consisting of points $x_1, x_2, ..., x_n$, the following functional representation

$$F(x_1, x_2, ..., x_n; \phi; \phi_{x_1}, \phi_{x_2}, ..., \phi_{x_n}; \phi_{x_1 x_1}, \phi_{x_1 x_2}, ... \phi_{x_1 x_n}, ...,) = 0,$$

$$x_1, x_2, ..., x_n (\equiv x) \in \Omega \tag{2.1}$$

furnishes a PDE with respect to an unknown function $\phi(x)$. The order of the highest derivative occurring in (2.1) determines the order of the PDE.

Introducing a multi-index notation an m-th order PDE can be presented in the form

$$F(x; \phi; D^m \phi, D^{m-1} \phi, ..., D\phi) = 0, \quad x \in \Omega \tag{2.2}$$

in which the differential operator D^m stands for the successive partial derivatives

$$D^m \equiv \frac{\partial_1^{m_1}}{\partial x_1^{m_1}} \cdots \frac{\partial_n^{m_n}}{\partial x_n^{m_n}} \equiv \frac{\partial^{|m|}}{\partial x_1^{m_1} \ldots \partial x_n^{m_n}}, \qquad x^m = x_1^{m_1} \ldots x_n^{m_n}$$

where $m_1, ..., m_n$ are non-negative integers and $|m| = m_1 + ... + m_n$. As a typical example, if $m = (4, 1, 2)$ then $|m| = 4 + 1 + 2 = 7$, $m! = 4!1!2! = 48$ and $D^m = \frac{\partial^7}{\partial^4 x_1 \partial x_2 \partial^2 x_3} = \partial_{x_1}^4 \partial_{x_2} \partial_{x_3}^2$.

A PDE expressed as

$$L\phi = \rho(x, y)$$

is linear when the operator L involves only the first powers of the partial differential operators

$$\sum_{|k| \leq m} p_k(x) D^k \phi(x) = \rho(x)$$

where the coefficient functions $p_k(x)$'s are assumed to be known. As already noted earlier, a inhomogeneous PDE is characterized by the non-vanishing of the term $\rho(x, y)$. Otherwise it is a homogeneous equation.

An arrangement of (2.2) in the form

$$\sum_{|k|=m} p_k(x)(D^{m-1}\phi, ..., D\phi, \phi, x) D^k \phi + p_0(x)(D^{m-1}\phi, ..., D\phi, \phi, x) = 0$$

defines a quasi-linear equation where it is clear that the highest-derivatives appear only linearly, whatever the functional nature of their coefficients.

A semi-linear equation, on the other hand, has the feature

$$\sum_{|k|=m} p_k(x) D^k \phi + p_0(x)(D^{m-1}\phi, ..., D\phi, \phi, x) D^k \phi + p_0(D^{m-1}\phi, ..., D\phi, \phi, x) = 0$$

which has a quasi-linear structure but the coefficients of the highest-derivatives are restricted to be functions of the multi-index vector x only. Typical examples of a quasi-linear and semi-linear PDE have already been given in chapter 1.

Any deviation from a linear character of a PDE would classify the equation as a nonlinear equation. The second order Monge-Ampére PDE

$$\det \left| \frac{\partial^2 \phi}{\partial x_i \partial x_j} \right| + \sum_{i,j=1} B_{ij}(x, \phi, D\phi) \frac{\partial^2 \phi}{\partial x_i \partial x_j} + C(x, \phi, D\phi) = 0$$

where the coefficients B_{ij} and C are functions of x, ϕ and $D\phi$ with $D\phi = (D_1\phi, ..., D_n\phi)$, $D_i = \frac{\partial}{\partial x_i}, i = 1, 2, ..., n$, is an example[1] of a nonlinear PDE. This equation has much relevance in optical transport problems. However, its severe nonlinear character makes an analytic determination of its solutions rather a formidable task[2]. To seek a solution for it one often has to take recourse to a numerical treatment.

Expressed in terms of two independent variables x and y, the Monge-Ampére PDE takes the simplified form

$$\phi_{xx}\phi_{yy} - \phi_{xy}^2 = \xi\phi_{xx} + 2\eta\phi_{xy} + \zeta\phi_{yy} + \chi$$

where ξ, η, ζ and χ are functions of x, y, ϕ along with the first derivatives of ϕ: ϕ_x and ϕ_y.

In the rest of this chapter we will be interested mostly in the linear PDEs of two independent variables. As we already know from Chapter 1 the operator L for a first order linear PDE has the form

$$L\phi \equiv a(x,y)\phi_x + b(x,y)\phi_y + c(x,y)\phi \tag{2.3}$$

the coefficients a, b, c being assumed to be continuous differentiable functions of x and y while for the second order linear PDE L has a general representation

$$L\phi \equiv A(x,y)\phi_{xx} + 2B(x,y)\phi_{xy} + C(x,y)\phi_{yy} + D(x,y)\phi_x + E(x,y)\phi_y + F(x,y)\phi \tag{2.4}$$

where A, B, C, D, E and F are assumed to be sufficiently smooth and continuously differentiable functions of x and y in a certain given domain.

2.2 Reduction to normal or canonical form

We focus on the second order linear PDE in two independent variables x and y where L is defined by (2.4). The second order linear PDEs can be classified much in the same way as is done when dealing with different types of conics in two-dimensional coordinate geometry. Recall that the three types of

---

[1]C. E. Gutiérrez, The Monge–Ampére Equation, Progress in Nonlinear Differential Equations and Their Applications, vol. 44, Birkhäuser, Basel, 2001.

[2]See, for instance, K. Brix, Y. Hafizogullari and A. Platen, Solving the Monge–Ampére equations for the inverse reflector problems, arXiv:1404.7821 /math.NA

conics are the hyperbola, ellipse and parabola according to whether the sign of the underlying discriminant[3] is positive, negative or zero.

The relevant discriminant for the second order linear PDEs is the quantity J defined by

$$J(x,y) = B^2(x,y) - A(x,y)C(x,y) \qquad (2.5)$$

A second order linear PDE $L\phi = 0$ can thus be classified to be of hyperbolic type, elliptic type or parabolic type in an arbitrary chosen neighbourhood in the xy-plane according to the signs of J. The different types of classifications go as follows:

$$B^2 - AC < 0 : \text{elliptic}$$
$$B^2 - A > 0 : \text{hyperbolic}$$
$$B^2 - AC = 0 : \text{parabolic} \qquad (2.6)$$

It is evident that with A,B,C depending on the independent variables x and y, the same PDE can reflect different characters for different neighbourhoods in the xy-plane. We already encountered, in the previous chapter, the three simplest representative equations of the elliptic, hyperbolic and parabolic equations as respectively the Laplace's equation, the wave equation and the heat conduction equation. One might be curious to know about the character of the nonlinear Monge-Ampére equation involving two independent variables. We just mention in passing that its type[4] depends on the sign of the quantity $Q \equiv \xi\zeta + \eta^2 + \chi$. Specifically, if $Q > 0$ then it is elliptic, if $Q < 0$ it is hyperbolic and if $Q = 0$ it is parabolic.

We now take up the issue of the reduction of a PDE to its normal form or canonical form. To this end, we ask the question that, given a regular curve γ in the xy-plane which is referred to as a boundary curve, if both ϕ and its derivative $\frac{\partial \phi}{\partial n}$ in a direction normal to the curve are prescribed, can we find a solution of $L(\phi) = 0$ near the curve?

To proceed with this aim we note, first of all, that if $\phi(x,y)$ is known along the boundary curve (or on some portion of it) where both x and y are parametrized by the same parameter s, where we identify s as the arc length measuring the distance along the curve (see Figure 2.1), then the derivative $\frac{\partial \phi}{\partial s}$ in the tangent direction to the boundary curve is known too. From $\frac{\partial \phi}{\partial n}$ and $\frac{\partial \phi}{\partial s}$ we can therefore determine both the partial derivatives ϕ_x and ϕ_y as explained below.

---

[3]S.L.Loney, The Elements of Coordinate Geometry, Arihant Publications (2016).
[4]Encyclopedia of Mathematics, Springer, The European Mathematical Society.

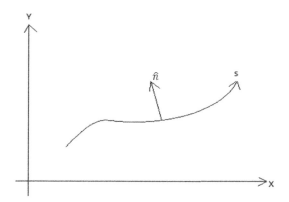

FIGURE 2.1: The boundary curve γ parametrized by s.

From the the theory of vector analysis, a vector having components $(\frac{dx}{ds}, \frac{dy}{ds})$ is tangent to the boundary curve which has a perpendicular vector having components $(-\frac{dy}{ds}, \frac{dx}{ds})$. We can therefore write

$$\frac{\partial \phi}{\partial n} = (\phi_x, \phi_y).(-\frac{dy}{ds}, \frac{dx}{ds}) = -\phi_x \frac{dy}{ds} + \phi_y \frac{dx}{ds} \tag{2.7}$$

Also

$$\frac{\partial \phi}{\partial s} = \phi_x \frac{dx}{ds} + \phi_y \frac{dy}{ds} \tag{2.8}$$

where the quantities ϕ_x and ϕ_y are evaluated corresponding to $x = x(s)$ and $y = y(s)$. From the above equations we therefore solve for ϕ_x and ϕ_y at some required point (x_0, y_0) in terms of the derivatives $\frac{\partial \phi}{\partial s}$ and $\frac{\partial \phi}{\partial n}$ known there.

Turning to the derivatives of ϕ_x and ϕ_y we use the chain rule of calculus to write down

$$\frac{d}{ds}\phi_x = \phi_{xx}\frac{dx}{ds} + \phi_{xy}\frac{dy}{ds} \tag{2.9}$$

$$\frac{d}{ds}\phi_y = \phi_{xy}\frac{dx}{ds} + \phi_{yy}\frac{dy}{ds} \tag{2.10}$$

From the above two equations and the given $L(\phi) = 0$ in which the second order partial derivatives also appear, it is evident that after arranging them appropriately we can always solve for the quantities $\phi_{xx}, \phi_{xy}, \phi_{yy}$ uniquely unless the determinant of the coefficient vanishes:

$$\triangle \equiv \begin{vmatrix} \frac{dx}{ds} & \frac{dy}{ds} & 0 \\ 0 & \frac{dx}{ds} & \frac{dy}{ds} \\ A & B & C \end{vmatrix} = 0$$

$\triangle$ when expanded yields

$$A\,dy^2 - 2B\,dx\,dy + C\,dx^2 = 0 \rightarrow A(\frac{dy}{dx})^2 - 2B\frac{dy}{dx} + C = 0, \quad A \neq 0 \quad (2.11)$$

It yields two families of curves each with its own constant of integration. Indeed there exist two directions given by the corresponding values of $\frac{dy}{dx}$ for which two families of curves (real or conjugate complex) exist. Such curves are called the characteristic curves. Of course on such curves (or lines) the quantity $\triangle$ vanishes i.e. $\triangle = 0$ and so as a necessary condition of solvability of the second derivatives $\phi_{xx}, \phi_{xy}, \phi_{yy}$ at a particular point the boundary curve must nowhere be tangent to a characteristic at that point. It should be remarked that if $A = 0$ but $C \neq 0$ then we should go for the equation in $\frac{dx}{dy}$ (as worked out in Example 2.5). Of course if both $A = 0$ and $C = 0$ we already are in a reduced form. The relevance of the characteristic curves to the propagation of singularities is discussed in Example 2.1 in the context of the one-dimensional wave equation.

Thus to know the values of the second partial derivatives at a particular point it is necessary to demand that $\triangle \neq 0$. To find the third order and higher partial derivatives we have to carry out repeated operations of further differentiations. Assuming the Taylor series to be convergent we can project the solution of $\phi(x, y)$ about the point (x_0, y_0) in the manner

$$\phi(x, y) = \phi_0 + [(x - x_0)(\phi_x)_0 + (y - y_0)(\phi_y)_0] +$$
$$+ \frac{1}{2!}[(x - x_0)^2(\phi_{xx})_0 + 2(x - x_0)(y - y_0)(\phi_{xy})_0 + (y - y_0)^2(\phi_{yy})_0] + \cdots$$
$$(2.12)$$

It is worthwhile repeating that prescribing data on a characteristic curve (on which $\triangle = 0$) has generally no meaning since it does not lead to a viable solution. It is only when the curve γ is non-characteristic ($\triangle \neq 0$) that the data on it uniquely determine the second and higher-derivatives appearing in the Taylor expansion of $\phi(x, y)$ about some suitable point of expansion.

We now examine the following examples.

Example 2.1

Consider the one-dimensional wave equation

$$\phi_{xx} - \frac{1}{c^2}\phi_{tt} = 0$$

Here $A = 1$, $B = 0$ and $C = -\frac{1}{c^2}$. The characteristics therefore obey the equations

$$\frac{dx}{dt} = \pm c$$

and thus turn out to be straight lines

$$x - ct = \xi = \text{constant}$$

$$x + ct = \eta = \text{constant}$$

A sketch of the family of lines is shown. See Figure 2.2.

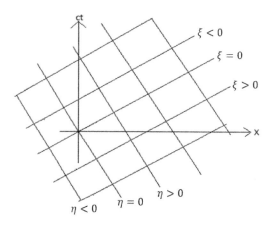

FIGURE 2.2: The characteristics lines.

The general solution of the wave equation is given by the sum of left and right travelling wave solutions

$$\phi(x,t) = f(x + ct) + g(x - ct)$$

If the initial values are assigned as

$$\phi(x,0) = 0, \quad \phi_t(x,0) = \delta(x)$$

where $\delta(x)$ is Dirac's delta function, then remembering that the derivative of the Heaviside function is the delta function (see equation (A.91) in Appendix A), the resultant solution which obey these conditions is easily worked out to be

$$\phi(x,t) = \frac{1}{2c}[H(x+ct) - H(x-ct)]$$

where H is the Heaviside function. The above solution of ϕ serves[5] as an interesting example of propagating singularity showing propagation to the left and right. In fact, for a hyperbolic equation every singularity (i.e. discontinuities of the boundary values) is continued along the characteristics and isolated singularities are not admitted. In contrast for an elliptic type of PDE the propagation of singularities can only be thought as being carried out in an imaginary domain.

Example 2.2

Consider the one-dimensional heat conduction equation

$$\phi_t = \alpha^2 \phi_{xx}$$

The characteristic curves are given by $dt^2 = 0$ or $t =$ constant which are parallel lines as shown the Figure 2.3.

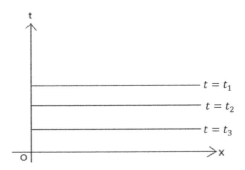

FIGURE 2.3: The characteristics lines.

As already touched upon in the previous chapter we can effectively use the characteristic variables to convert the PDE in (x, y) variables to a special

---

[5] Jeffrey Rauch, Lecture notes on Hyperbolic PDEs and geometrical optics, American Mathematical Society (2012).

form called the normal or canonical form. In the present context we solve the condition $\triangle = 0$ as furnished by (2.11) to obtain the pair of solutions

$$y_\pm = \int_x \frac{dy}{dx}\, dx = \int_x (\frac{B \pm \sqrt{B^2 - AC}}{A})\, dx \qquad (2.13)$$

These provide the two equations of characteristics which are functions of x and y when the given forms of A, B and C are substituted in (2.13) and integration is carried out.

Let us express the characteristic equations according to the prescriptions

$$\xi(x, y) = \text{constant}, \quad \eta(x, y) = \text{constant} \qquad (2.14)$$

Our task would be to transform the independent variables (x, y) in terms of the new ones (ξ, η). With a non-vanishing Jacobian of the transformatiom which reads

$$\bar{\triangle} \equiv \begin{vmatrix} \xi_x & \xi_y \\ \eta_x & \eta_y \end{vmatrix} \neq 0$$

a one-to-one relation between the old and new variables is assured.

The first and second order derivatives of ϕ can be transformed in terms of the variables ξ and η by employing the chain rule of partial derivatives. We find in this way for the first derivatives

$$\phi_x = \phi_\xi \xi_x + \phi_\eta \eta_x, \quad \phi_y = \phi_\xi \xi_y + \phi_\eta \eta_y \qquad (2.15)$$

and subsequently for the second derivatives the set of relations

$$\phi_{xx} = \phi_{\xi\xi}\xi_x^2 + 2\phi_{\xi\eta}\xi_x\eta_x + \phi_{\eta\eta}\eta_x^2 + \dots \qquad (2.16)$$

$$\phi_{xy} = \phi_{\xi\xi}\xi_x\xi_y + \phi_{\xi\eta}(\xi_x\eta_y + \xi_y\eta_x) + \phi_{\eta\eta}\eta_x\eta_y + \dots \qquad (2.17)$$

$$\phi_{yy} = \phi_{\xi\xi}\xi_y^2 + 2\phi_{\xi\eta}\xi_y\eta_y + \phi_{\eta\eta}\eta_y^2 + \dots \qquad (2.18)$$

where the dots in the right side of each equation represent the lower order derivative terms.

Hence, in terms of the variables (ξ, η), the second order partial derivative terms present in (2.4) go over to the form

$$A\phi_{xx} + 2B\phi_{xy} + C\phi_{yy} \rightarrow \tilde{A}\phi_{\xi\xi} + +2\tilde{B}\phi_{\xi\eta} + \tilde{C}\phi_{\eta\eta}$$

where the new coefficients $\tilde{A}$, $\tilde{B}$ and $\tilde{C}$ are given by

$$\tilde{A} = A\xi_x^2 + 2B\xi_x\xi_y + C\xi_y^2, \quad \tilde{B} = A\xi_x\eta_x + B(\xi_x\eta_y + \xi_y\eta_x) + C\xi_y\eta_y,$$
$$\tilde{C} = A\eta_x^2 + 2B\eta_x\eta_y + C\eta_y^2 \tag{2.19}$$

Notice that since

$$\tilde{B}^2 - \tilde{A}\tilde{C} = (\xi_x\eta_y - \xi_y\eta_x)^2(B^2 - AC) \tag{2.20}$$

the intrinsic character of the PDE is not changed under the transformation $(x, y) \to (\xi, \eta)$.

For the characteristic family given by $\xi(x, y) = $ constant we have

$$\xi_x dx + \xi_y dy = 0 \quad \to \quad \frac{dy}{dx} = -\frac{\xi_x}{\xi_y} \tag{2.21}$$

which transforms the condition (2.11) to

$$A\xi_x^2 + 2B\xi_x\xi_y + C\xi_y^2 = 0 \tag{2.22}$$

This implies from (2.19) that the coefficient $\tilde{A}$ of $\phi_{\xi\xi}$ vanishes.

Similarly for the other characteristic family given by $\eta(x, y) = $ constant one runs into the form

$$A\eta_x^2 + 2B\eta_x\eta_y + C\eta_y^2 = 0 \tag{2.23}$$

which points to the vanishing of the coefficient $\tilde{C}$ of $\phi_{\eta\eta}$ in (2.19).

Combining the above two results one arrives at the following canonical form of the hyperbolic equation

$$\phi_{\xi\eta} = f(\phi, \phi_\xi, \phi_\eta, \xi, \eta) \tag{2.24}$$

where f is an arbitrary function of its arguments. This form corresponds to the case $B \neq 0$ and $A = 0 = C$ which ensures that the hyperbolic condition $B^2 - AC > 0$ is automatically fulfilled.

On the other hand, the condition $B^2 - AC > 0$ is also met through having $B = 0$ and $A = -C$. The manifestation of this case comes through the transformations

$$\xi = x + y, \quad \eta = x - y \tag{2.25}$$

which provides a different canonical form of the hyperbolic equation namely,

$$\phi_{xx} - \phi_{yy} = g(\phi, \phi_x, \phi_y, x, y) \tag{2.26}$$

where g is the transformed function of its arguments.

We need not deal with the elliptic case separately for it follows from the hyperbolic case on employing the complex transformations

$$\xi = x + iy, \quad \eta = x - iy \tag{2.27}$$

Indeed (2.27) results in

$$\phi_{xx} + \phi_{yy} = h(\phi, \phi_x, \phi_y, x, y) \tag{2.28}$$

where h is the transformed function of its arguments. It is clear that in the elliptic case the characteristics are conjugate complex. Example 1.12 worked out in the previous chapter will help to clear up this point.

For the parabolic case, we set $\eta = x$ but keep ξ free. In such a situation the left side of the relation (2.22) becomes a perfect square i.e.

$$(A\xi_x + B\xi_y)^2 = 0 \tag{2.29}$$

Here we employed the parabolic condition $B^2 - AC = 0$.

Now with the help of the relations (2.16), (2.17) and (2.18) we find

$$A\phi_{xx} + 2B\phi_{xy} + C\phi_{yy} = 2\phi_{\xi\eta}(A\xi_x + B\xi_y) + A\phi_{\eta\eta} = A\phi_{\eta\eta} \tag{2.30}$$

where we have used (2.29). As a result we get the parabolic normal form

$$\phi_{\eta\eta} = G(\phi, \phi_\xi, \phi_\eta, \xi, \eta) \tag{2.31}$$

where G is an arbitrary function of its arguments.

The following examples serve to illustrate the reduction of a PDE to its normal form.

Example 2.3

Reduce the following PDE to its normal form

$$y\phi_{xx} + (x+y)\phi_{xy} + x\phi_{yy} = 0$$

Here $A = y$, $B = \frac{x+y}{2}$ and $C = x$. As a result

$$B^2 - AC = \frac{(x-y)^2}{4} > 0$$

Hence the equation is hyperbolic except on the line $x = y$ on which it is parabolic. In the following we focus on the hyperbolic case.

The characteristic curves are obtainable from (2.11) which gives

$$\frac{dy}{dx} = \frac{x}{y} \quad \text{or} \quad 1$$

They therefore correspond to the equations $x^2 - y^2 = $ constant and $x - y = $ constant. Let us set

$$\xi = x^2 - y^2, \quad \eta = x - y$$

Transforming to the variables ξ and η we find for the first derivatives ϕ_x and ϕ_y

$$\phi_x = 2x\phi_\xi + \phi_\eta, \quad \phi_y = -2y\phi_\xi - \phi_\eta$$

and for the second derivatives ϕ_{xx}, ϕ_{xy} and ϕ_{yy}

$$\phi_{xx} = 4x^2\phi_{\xi\xi} + 4x\phi_{\xi\eta} + \phi_{\eta\eta} - 2\phi_\xi$$
$$\phi_{xy} = -4xy\phi_{\xi\xi} - 2(x+y)\phi_{\xi\eta} - \phi_{\eta\eta}$$
$$\phi_{yy} = 4y^2\phi_{\xi\xi} + 4y\phi_{\xi\eta} + \phi_{\eta\eta} - 2\phi_\xi$$

On substitution of the above results the given equation is converted to

$$(x-y)^2\phi_{\xi\eta} + (x+y)\phi_\xi = 0$$

which implies that the normal form is

$$\eta^3\phi_{\xi\eta} + \xi\phi_\xi = 0$$

Example 2.4

Solve the PDE

$$\frac{\partial}{\partial y}(\phi_x + \phi) + 2x^2 y(\phi_x + \phi) = 0$$

by transforming it to the normal form.
    The given equation reads

$$\phi_{xy} + \phi_y + 2x^2 y(\phi_x + \phi) = 0$$

Here $A = 0$, $B = \frac{1}{2}$ and $C = 0$. Hence $B^2 - AC = \frac{1}{4} > 0$ and so the equation is hyperbolic. The equations of the characteristics are given by

$$\frac{dy}{dx} = 0, \quad \frac{dx}{dy} = 0$$

and imply that $y = $ constant and $x = $ constant are the characteristic lines. We therefore put

$$\xi = x = \text{constant}, \quad \eta = y = \text{constant}$$

Thus the given equation has the same form in terms of ξ and η variables:

$$\frac{\partial}{\partial \eta}(\phi_\xi + \phi) + 2\xi^2 \eta(\phi_\xi + \phi) = 0$$

Let us put

$$\zeta = \phi_\xi + \phi$$

which transforms the above equation to a rather simple representation

$$\frac{\partial \zeta}{\partial \eta} + 2\xi^2 \eta \zeta = 0$$

Integrating partially with respect to η gives

$$\zeta(\xi, \eta) = f(\xi)e^{-\xi^2 \eta^2}$$

where f is an arbitrary function of ξ.

    Transforming back the ϕ variable and integrating partially with respect to ξ we obtain

$$\phi(\xi, \eta) = e^{-\xi}[g(\eta) + \int f(\xi)e^{\xi - \xi^2 \eta^2} d\xi]$$

where g is an arbitrary function of η.

Example 2.5

Solve the following PDE by transforming it to the normal form

$$e^y \phi_{xy} - \phi_{yy} + \phi_y = 0$$

where it is given that at $y = 0$, $\phi_{xy} = -\frac{x^2}{2}$ and $\phi_y = -\sin x$.

Here $A = 0$, $B = \frac{e^y}{2}$ and $C = -1$. Since $A = 0$, as noted before, we need to consider the equation in $\frac{dx}{dy}$ which reads

$$A - 2B(\frac{dx}{dy}) + C(\frac{dx}{dy})^2 = 0$$

We thus have

$$\frac{dx}{dy}(-2B + C\frac{dx}{dy}) = 0$$

Plugging in the forms of B and C we find

$$\frac{dx}{dy} = 0, \quad \frac{dx}{dy} = -e^y$$

The first equation imples $x = $ constant and the second one on integration gives $x + e^y = $ constant. We therefore identify the ξ and η variables to be

$$\xi = e^y + x - 1, \quad \eta = x$$

where -1 has been inserted without any loss of generality.

It is an easy task to work out the first derivative conversions

$$\phi_x = \phi_\xi + \phi_\eta, \quad \phi_y = e^y \phi_\xi$$

and in consequence have the follwoing results for the second derivatives

$$\phi_{xx} = \phi_{\xi\xi} + 2\phi_{\xi\eta} + \phi_{\eta\eta}, \quad \phi_{xy} = e^y(\phi_{\xi\xi} + \phi_{\xi\eta}), \quad \phi_{yy} = e^y \phi_\xi + e^{2y}\phi_{\xi\xi}$$

When substituted in the given equation we are led to the normal form

$$\phi_{\xi\eta} = 0$$

Integrating easily we find the solution for ϕ to be

$$\phi(\xi, \eta) = f(\xi) + g(\eta)$$

where f and g are two arbitrary functions of their arguments.
In terms of the variables x and y we thus have

$$\phi(x, y) = f(e^y + x - 1) + g(x)$$

Employing the given conditions we can determine f and g. This gives the solution

$$\phi(x, y) = -\frac{x^2}{2} + \cos(e^y + x - 1) - \cos x$$

We conclude this section by making a few remarks on a fluid motion where we can categorize the flow in the following way. For a steady viscous flow it is elliptic whereas it is parabolic for the unsteady case. For an inviscid flow, the elliptic or hyperbolic nature depends on the Mach number being less than or greater than one. However, for the unsteady inviscid flow it is always hyperbolic.

As an illustration let us consider a two-dimensional potential equation of a steady, irrotational, inviscid flow[6] in terms of the velocity potential Φ

$$(1 - \frac{\Phi_x^2}{c^2})\Phi_{xx} - 2\frac{\Phi_x\Phi_y}{c^2}\Phi_{xy} + (1 - \frac{\Phi_y^2}{c^2})\Phi_{yy} = 0$$

where c is the local speed of sound, Φ_x and Φ_y are the x and y components of the flow velocity and the Mach number is $M = \frac{\sqrt{\Phi_x^2 + \Phi_y^2}}{c}$. The square of the discriminant of the equation is easily seen to be

$$\frac{\Phi_x^2\Phi_y^2}{c^4} - (1 - \frac{\Phi_x^2}{c^2})(1 - \frac{\Phi_y^2}{c^2}) = \frac{\Phi_x^2}{c^2} + \frac{\Phi_y^2}{c^2} - 1 = M^2 - 1$$

We therefore find that for a subsonic flow, when $M < 1$, the equation is elliptic while for a supersonic flow, when $M > 1$, the equation is hyperbolic. The parabolic character occurs when the sonic condition holds.

---

[6]R. Vos and S. Farokhi, Introduction to Transonic Aerodynamics, Fluid Mechanics and its Applications, Springer Science+Business Media, Dordrecht, 2015.

2.3 Boundary and initial value problems

(a) Different types of boundary and initial value problems

An extended form of elliptic PDE that includes as particular cases the Poisson equation, Laplace equation and Helmholtz equation is given by

$$-\nabla \cdot (\alpha \nabla \phi) + \beta \phi = f \qquad (2.32)$$

where α, β and f are functions of the space variable $\vec{r}$. In fact, (2.32) reduces to the Poisson's form when $\alpha = 1$, $\beta = 0$ and $f = 4\pi\rho$ while Laplace's equation follows when $\alpha = 1$, $\beta = 0$ and $f = 0$. The case when $\alpha = 1$, $f = 0$ and $\beta = -k^2$ corresponds to the Helmholtz's form.

In a general way, let us consider a domain Ω specified by its boundary $\partial\Omega$ such that a solution ϕ of (2.32) has continuous second order derivatives in Ω and is also continuous in $\bar{\Omega}$, the closure of Ω, with the following condition being valid on $\partial\Omega$ i.e.

$$\partial\Omega : a\phi + b\frac{\partial \phi}{\partial n} = v \qquad (2.33)$$

where a, b and v are given continuous functions on $\partial\Omega$ and the derivative in (2.33) is along the outward drawn normal to $\partial\Omega$. We read off from (2.33) three distinct boundary conditions as given by the following possibilities

(i) If $a = 1$ and $b = 0$ then we have

$$\partial\Omega : \phi = v \qquad (2.34)$$

It is referred to as the boundary condition of the first kind accompanying the boundary value problem of the first kind.

(ii) If $a = 0$ and $b = 1$ then

$$\partial\Omega : \frac{\partial \phi}{\partial n} = v \qquad (2.35)$$

It is referred to as the boundary condition of the second kind accompanying the boundary value problem of the second kind.

(ii) If $a \geq 0$ and $b = 1$ which implies

$$\partial\Omega : a\phi + \frac{\partial\phi}{\partial n} = v \tag{2.36}$$

It is referred to as the boundary condition of the third kind[7] accompanying the boundary value problem of the third kind.

Two standard types of boundary value problems that one meets in solving physical problems are those which go by the classifications of Dirichlet and the Neumann types. In the Dirichlet's case we need to find a solution ϕ which has continuous second order derivatives in Ω and also continuous in $\bar{\Omega}$. We express this problem by

$$\Omega : \nabla^2\phi = f, \quad \partial\Omega : \phi = g \tag{2.37}$$

where g is a continuous function.

In the Neumann's case we have to find a solution ϕ which has continuous second order derivatives in Ω and also continuous in $\bar{\Omega}$. We express it as

$$\Omega : \nabla^2\phi = f, \quad \partial\Omega : \frac{\partial\phi}{\partial n} = h \tag{2.38}$$

where h is a continuous function.

As opposed to an elliptic equation which addresses a boundary value problem, the hyperbolic case calls for the specification of initial conditions. The prototype of a hyperbolic PDE is the wave equation which in three dimensions reads

$$\nabla^2\phi - \frac{1}{c^2}\frac{\partial^2\phi}{\partial t^2} = 0 \tag{2.39}$$

In a standard initial value problem, the dependent function and its time derivative are specified on the boundary $\partial\Omega$ of some region Ω at some initial time. For instance, in a three-dimensional region (x, y, z) it is given that

$$\phi(x, y, z, 0) = f(x, y, z), \quad \phi_t(x, y, z, 0) = g(x, y, z), \quad \text{on} \quad \partial\Omega \tag{2.40}$$

As we shall see later these conditions determine a unique solution to the wave equation.

---

[7]Such a boundary condition is also referred to as Robin's type or mixed type.

An initial value problem as stated above defines a Cauchy problem for a second order linear PDE. As should be clear the related hypersurface in such a case should nowhere be a characteristic surface for the solvability of the Cauchy problem.

(b) Applications

We now turn to the two-dimensional Laplace's equation

$$\phi_{xx} + \phi_{yy} = 0 \tag{2.41}$$

Its solution is sought in the separable form

$$\phi(x, y) = X(x)Y(y) \tag{2.42}$$

Substitution in the Laplace equation gives

$$\frac{X''}{X} = -\frac{Y''}{Y} = k \tag{2.43}$$

where k is a separation constant arising due to one side being a function of x while the other is a function of y and dashes refer to derivatives with respect to the relevant variable.

We can distinguish three cases of the solutions

$$
\begin{aligned}
(i) \quad & k = 0 \Rightarrow \phi(x, y) = (ax + b)(cx + d) \\
(ii) \quad & k = p^2 > 0 \Rightarrow \phi(x, y) = (r\,e^{px} + f\,e^{-px})(g\,\cos(py) + h\,\sin(py)) \\
(iii) \quad & k = p^2 < 0 \Rightarrow \phi(x, y) = (j\,\cos(px) + s\,\sin(px))(l\,e^{py} + m\,e^{-py})
\end{aligned}
$$

where a, b, c, d, r, f, g, h, j, s, l, m are arbitrary constants.

With this little background let us first address the Dirichlet problem on a rectangular domain.

Dirichlet problem on a rectangular domain: $0 \leq x \leq a$, $0 \leq y \leq b$

Let us solve the two-dimensional Laplace's equation (2.41) by imposing the boundary conditions:

$$\phi(a, y) = 0 = \phi(x, b), \quad \phi(0, y) = 0, \quad \phi(x, 0) = f(x) \tag{2.44}$$

In this problem the case (iii) above is relevant and so on applying the conditions $\phi(0,y) = 0$ and $\phi(a,y) = 0$ we are led to

$$\phi(0,y) = 0 \Rightarrow j = 0$$
$$\phi(a,y) = 0 \Rightarrow s \neq 0, \quad \sin(pa) = 0 \Rightarrow p = \frac{n\pi}{a} \quad (n = 1, 2, ...)$$

Thus $\phi(x,y)$ takes the form

$$\phi(x,y) = \sum_{n=1}^{\infty} A_n \sin \frac{n\pi x}{a} (l_n e^{\frac{n\pi y}{a}} + m_n e^{-\frac{n\pi y}{a}}) \tag{2.45}$$

where A_n is an overall coefficient and l_n, m_n, $n = 1, 2, ...$ are constants.

We now exploit the condition $\phi(x,b) = 0$ to have

$$\phi(x,b) = 0 \quad \Rightarrow \quad m_n = -l_n \frac{e^{\frac{n\pi b}{a}}}{e^{-\frac{n\pi b}{a}}} \tag{2.46}$$

As a result we can express $\phi(x,y)$ as

$$\phi(x,y) = \sum_{n=1}^{\infty} A_n \sin \frac{n\pi x}{a} \sinh[\frac{n\pi}{a}(y-b)] \tag{2.47}$$

where the coefficient A_n can be determined from the remaining condition $\phi(x,0) = f(x)$:

$$\phi(x,0) = f(x) \Rightarrow \sum_{n=1}^{\infty} A_n \sin \frac{n\pi x}{a} \sinh(-\frac{n\pi b}{a}) = f(x) \tag{2.48}$$

On inversion we readily find

$$A_n = \frac{2}{a} \frac{1}{\sinh(-\frac{n\pi b}{a})} \int_0^a f(x) \sin(\frac{n\pi x}{a}) \, dx \tag{2.49}$$

The final form of $\phi(x,y)$ satisfying all the given boundary conditions is given by (2.47) where for A_n the integral provided by (2.49) holds.

Next we look at the analogous problem for the Neumann problem in a rectangular domain.

Neumann problem on a rectangular domain: $0 \leq x \leq a$, $0 \leq y \leq b$

Here the Laplace's equation (2.41) is solved subject to the boundary conditions

$$\frac{\partial \phi(0,y)}{\partial x} = 0, \quad \frac{\partial \phi(a,y)}{\partial x} = 0; \quad \frac{\partial \phi(x,0)}{\partial y} = 0, \quad \frac{\partial \phi(x,b)}{\partial y} = f(x) \quad (2.50)$$

Here again case (iii) is relevant. Proceeding as in the previous case we find for $\phi(x,y)$ the form

$$\phi(x,y) = B_0 + \sum_{n=1}^{\infty} B_n \cos \frac{n\pi x}{a} \cosh \frac{n\pi y}{a} \quad (2.51)$$

where B_n is given by the integral

$$B_n = \frac{2}{n\pi} \frac{1}{\sinh(\frac{n\pi b}{a})} \int_0^a f(x) \cos \frac{n\pi x}{a} \, dx, \quad (n = 1, 2, ...) \quad (2.52)$$

It may be remarked that in an exterior boundary value problem for an elliptic type of equation, the problem is to find a solution in the region Ω' exterior to Ω which is unbounded. In such a case, ϕ has to satisfy a condition at infinity controlling the asymptotic behaviour of the solution.

Cauchy initial value problem on an interval: $0 \leq x \leq l$, $t > 0$

We consider the one-dimensional wave equation

$$\phi_{tt} - c^2 \phi_{xx} = 0, \quad 0 \leq x \leq l, \quad t > 0 \quad (2.53)$$

subject to the following Cauchy data

$$\phi(x,0) = g(x), \quad \phi_t(x,0) = h(x), \quad 0 \leq x \leq l \quad (2.54)$$

We also implement a set of boundary conditions

$$\phi(0,t) = 0, \quad \phi(l,t) = 0, \quad t \geq 0 \quad (2.55)$$

Let us first of all write the general solution in the form

$$\phi(x,t) = \frac{A_0(t)}{2} + \sum_{n=1}^{\infty}[A_n(t)\cos\frac{n\pi x}{l} + B_n(t)\sin\frac{n\pi x}{l}] \qquad (2.56)$$

where A_n and B_n are appropriate coefficients. Because of the boundary condition $\phi(0,t) = 0$, the quantities $A_n, n = 0, 1, 2, ...$ have to vanish leaving us with the representation

$$\phi(x,t) = \sum_{n=1}^{\infty} B_n(t)\sin\frac{n\pi x}{l} \qquad (2.57)$$

In it the other boundary condition $\phi(l,t) = 0$ automatically holds.

Substituting in the given equation we therefore find that B_n's obey the differential equation

$$\ddot{B}_n(t) + (\frac{n\pi c}{l})^2 B_n(t) = 0 \qquad (2.58)$$

where the overhead dots denote derivatives with respect to t. The general solution of B_n turns out to be

$$B_n(t) = c_n\sin\frac{n\pi ct}{l} + d_n\cos\frac{n\pi ct}{l} \qquad (2.59)$$

where c_n and d_n are arbitrary constants to be determined from suitable inputs. Towards this end we exploit the initial conditions:

$$g(x) = \phi(x,0) = \sum_{n=1}^{\infty} B_n(0)\sin\frac{n\pi x}{l} = \sum_{n=1}^{\infty} d_n\sin\frac{n\pi x}{l} \qquad (2.60)$$

$$h(x) = \phi_t(x,0) = \sum_{n=1}^{\infty} \dot{B}_n(0)\sin\frac{n\pi x}{l} = \sum_{n=1}^{\infty} c_n\frac{n\pi c}{l}\sin\frac{n\pi x}{l} \qquad (2.61)$$

We now multiply both sides of the above equations by $\sin\frac{m\pi x}{l}$ and integrate between 0 and l. Then by orthogonality

$$\int_0^l g(x)\sin\frac{m\pi x}{l}dx = \int_0^l \sum_{n=1}^{\infty} d_n\sin\frac{n\pi x}{l}\sin\frac{m\pi x}{l}dx = d_m(\frac{l}{2}) \qquad (2.62)$$

$$\int_0^l h(x) \sin \frac{m\pi x}{l} dx = \int_0^l \sum_{n=1}^{\infty} c_n \frac{n\pi c}{l} \sin \frac{n\pi x}{l} \sin \frac{m\pi x}{l} dx = c_m\left(\frac{m\pi c}{2}\right) \quad (2.63)$$

where we have used $\int_0^l \sin \frac{n\pi x}{l} \sin \frac{m\pi x}{l} dx = \frac{l}{2}$ if $n = m$ but 0 if $n \neq m$.

Hence the constants c_n and d_n are

$$d_n = \frac{2}{l} \int_0^l g(x) \sin \frac{n\pi x}{l} dx, \quad c_n = \frac{2}{n\pi c} \int_0^l h(x) \sin \frac{n\pi x}{l} dx \quad (2.64)$$

With the knowledge of c_n and d_n, B_n is known from (2.59) which in turn gives $\phi(x,t)$ from (2.57).

(c) Well-posedness

A PDE is said to be well-posed if a solution to it exists (i.e. at least one), it is unique (i.e. at most one) and that it depends continuously on the data (i.e. a small change in the initial data produces a correspondingly small change in the solution). For the existence and uniqueness of the solution appropriate initial and/or boundary conditions need to be imposed.

Take for instance the Laplace's equation defined over a rectangular region (a, b)

$$\phi_{xx} + \phi_{yy} = 0, \quad 0 < x < a, \quad 0 < y < b$$

subject to obeying the boundary conditions

$$\phi(x,0) = x^2, \quad \phi(x,b) = x^2 - b^2, \quad \phi(0,y) = -y^2, \quad \phi(a,y) = a^2 - y^2$$

It is straightforward to see that a solution exists in the form

$$\phi(x,y) = x^2 - y^2$$

which satisfies the given boundary conditions and is unique as well. One cannot afford to do away with any of the boundary conditions as that would lead to multiple solutions and the problem would be rendered ill-posed.

A few comments on the appropriateness of the boundary conditions are in order by focusing on the following Hadamard's example.

Hadamard example

It is evident that the solution of the PDE

$$\phi_{xx}(x,t) + \phi_{tt}(x,t) = 0 \qquad -\infty < x < \infty, \quad t > 0$$

subject to the homogeneous conditions

$$\begin{aligned}\phi(x,0) &= 0 & -\infty < x < \infty, \\ \phi_t(x,0) &= 0 & -\infty < x < \infty\end{aligned}$$

is given by

$$\phi(x,t) = 0$$

In contrast the solution of the same PDE i.e.

$$\psi_{xx}(x,t) + \psi_{tt}(x,t) = 0, \qquad -\infty < x < \infty, \quad t > 0$$

which is elliptic and being subjected to Cauchy-like initial conditions

$$\begin{aligned}\psi(x,0) &= 0 & -\infty < x < \infty \\ \psi_t(x,0) &= \epsilon \sin(\frac{x}{\epsilon}) & -\infty < x < \infty\end{aligned}$$

where $\epsilon > 0$ is a small quantity, has the solution

$$\phi(x,t) = \epsilon^2 \sin(\frac{x}{\epsilon}) \sinh(\frac{t}{\epsilon})$$

It follows that while

$$|\psi_t(0) - \phi_t(0)| = \epsilon \to 0 \quad \text{as} \quad \epsilon \to 0$$

the difference of the time-dependent solutions behaves

$$|\psi(t) - \phi(t)| = \epsilon^2 |\sinh(\frac{t}{\epsilon})| \to \infty \quad \text{as} \quad \epsilon \to 0$$

i.e. exponentially blows up. We thus see that a small change in the initial data produces a massive change in the solution. We conclude that there is a failure to depend continuously on the supplied data and the problem is ill-posed.

We remind the readers that if we had replaced the Laplace equation by the hyperbolic wave equation

$$\phi_{tt}(x,t) - \phi_{xx}(x,t) = 0 \qquad -\infty < x < \infty, \quad t > 0$$

and subjected it to the homogeneous conditions

$$\begin{aligned}\phi(x,0) &= 0 & -\infty < x < \infty \\ \phi_t(x,0) &= 0 & -\infty < x < \infty\end{aligned}$$

a trivial solution would be

$$\phi(x,t) = 0$$

However, the same form of the PDE i.e.

$$\psi_{xx}(x,t) - \psi_{tt}(x,t) = 0 \qquad -\infty < x < \infty, \quad t > 0$$

which is now hyperbolic be subjected to Cauchy-like conditions

$$\psi(x,0) \;=\; 0 \qquad\qquad -\infty < x < \infty$$
$$\psi_t(x,0) \;=\; \epsilon \sin\left(\frac{x}{\epsilon}\right) \qquad -\infty < x < \infty$$

would result in the solution

$$\psi(x,t) = \epsilon^2 \sin\left(\frac{x}{\epsilon}\right) \sin\left(\frac{t}{\epsilon}\right)$$

Comparing with previous case we find while

$$|\psi_t(0) - \phi_t(0)| = \epsilon \to 0 \quad \text{as} \quad \epsilon \to 0$$

the difference of the time-dependent solutions behaves

$$|\psi(t) - \phi(t)| = \epsilon^2 |\sin\left(\frac{t}{\epsilon}\right)| \le \epsilon^2 \quad \text{as} \quad \epsilon \to 0$$

We thus infer that if we made a small change in the initial data it would reflect a corresponding small change in the solution as well. Thus unlike the previous case the problem here is well-posed.

We conclude this section explaining the perspective of the well known Cauchy-Kowalevski theorem that gives a sufficient condition for the local solvability of the Cauchy problem in some neighbourhood of an initial point. We consider a Cauchy system which consists of a set of $(n+1)$- variables distinguished by a set of n spatial coordinates $x_1, x_2, ..., x_n$ and additionally a time parameter t. In terms of these we look at the following set of PDEs in terms of the functions $\phi_1, \phi_2, ..., \phi_N$ written in a compact but self-evident form

$$\frac{\partial^{n_i} \phi_i}{\partial t^{n_i}} = \Phi_i(x_1, x_2, ..., x_n; t; \phi_1, \phi_2, ..., \phi_N; D^k \phi_j; ...), \quad 1 \le i, j \le N \quad (2.65)$$

where in the left side there are partial derivatives of t of order n_i operating on $\phi_1, \phi_2, ..., \phi_N$ each of which is a function of the variables $x_1, x_2, ..., x_n$ and in the right side the Φ_i's are arbitrary functions depending on the variables $x_1, x_2, ..., x_n$, the time t along with $\phi_1, \phi_2, ..., \phi_N$. Adopting the multi-indexed notation introduced earlier in this chapter, the differential operator D^k denotes

$$D^k \equiv \frac{\partial^{|k|}}{\partial t^{k_0} \partial x_1^{k_1} ... \partial x_n^{k_n}} \tag{2.66}$$

where

$$|k| = k_0 + k_1 + ... + k_n \leq n_j, \quad k_0 < n_j \tag{2.67}$$

and that D^k involves involves the partial derivatives of t as well as of the variables $x_1, x_2, ..., x_n$ as shown by the respective superscripts.

An an example of a Cauchy system we may think of the following PDE[8] in two variables x and y

$$\frac{\partial^2 \phi}{\partial t^2} = t + x^2 - y^2 + x\frac{\partial \phi}{\partial x} + 2\frac{\partial^2 \phi}{\partial t \partial x} - \frac{\partial^2 \phi}{\partial x \partial y} + \frac{\partial^2 \phi}{\partial x^2} + \frac{\partial^2 \phi}{\partial y^2} \tag{2.68}$$

One can see that in this scheme we have just one equation which means $N = 1$, $n = 2$ because the number of variables are three t, x and y, the order of the partial-time derivative in the left side being two implies $n_i = 2$. The highest derivative in the right side of the space variables is also two while the highest derivative of time there is one.

The following coupled system also explains the notations

$$\frac{\partial^3 \phi_1}{\partial t^3} = \frac{\partial^2 \phi_1}{\partial x_1^2} - \frac{\partial^3 \phi_2}{\partial x_1^2 \partial x_2} + x_1 x_2 \frac{\partial^3 \phi_1}{\partial x_1 \partial x_2} + t^2 x_1 \tag{2.69}$$

$$\frac{\partial^2 \phi_2}{\partial t^2} = \frac{\partial^2 \phi_1}{\partial x_2^2} + \frac{\partial^2 \phi_2}{\partial x_1 \partial t} - \frac{\partial \phi_1}{\partial t} - \frac{\partial \phi_2}{\partial t} \tag{2.70}$$

Here it is easy to identify that $N = 2, n = 2$ and $n_1 = 3, n_2 = 2$.

Coming back to the set of N PDEs at hand, let us suppose that they are subjected to the initial conditions at $t = t_0$

$$\frac{\partial^k \phi_i}{\partial t^k}|_{t=t_0} = f_{i,k}(x_1, x_2, ..., x_n), \quad i = 1, 2, ..., N \tag{2.71}$$

At $k = 0$ we have the zeroth order partial derivative $\frac{\partial^0 \phi_i}{\partial t^0}|_{t=t_0}$ which is taken as the function of itself at $(t_0, x_1, x_2, ..., x_n)$.

Suppose now that the arguments of the functions $\Phi_1, \Phi_2, ..., \Phi_n$ are analytic in an open region containing the point $(t_0, x_1^0, x_2^0, ...x_n^0)$, the arguments reading explicitly

---

[8]P. V. O'Neil, Beginning Partial Differential Equations (1969), John Wiley and Sons, Inc., USA.

$$(x_1^0, x_2^0, ..., x_n^0; t_0, ..., [D^{k-k_0}\phi_{i,k^0}]_{k_1=x_1^0,...,k_n=x_n^0}) \qquad (2.72)$$

and that around an open region of $(x_1^0, x_2^0, ..., x_n^0)$ the functions $f_{i,k}$ are analytic too, then the Cauchy-Kowalevski theorem states that the initial-value problem for the Cauchy system, as defined by (2.66) together with the conditions (2.72), admits of exactly one unique solution which is analytic around $(x_1^0, x_2^0, ..., x_n^0; t_0)$.

2.4   Insights from classical mechanics

Let us consider the motion[9] of a particle in one-dimension. We discuss two cases of the force acting on the particle.

- Elastic force $-kx$ ($k > 0$)

The equation of motion can be written as

$$\dot{p} = -kx \qquad \text{where} \quad p = m\dot{x}$$

The general solution is

$$x = a \, \cos(\omega t + \phi) \qquad p = -m\omega a \, \sin(\omega t + \phi)$$

where $\omega = \sqrt{\frac{k}{m}}$.

In the (x, p) phase plane the representative point traces out an ellipse of semi-axes $(\sqrt{\frac{2E}{k}}, \sqrt{2mE})$, E being the energy. The ellipse is a closed curve. See Figure 2.4. In the theory of an elliptic PDE we already know that for a well meaning solution to exist we generally have either a Dirichlet or Neumann condition to be specified on the closed boundary.

- Repulsive force $+kx$ ($k > 0$)

Here we have two sets of solutions :

$$x \;=\; \pm a \, \cosh(\omega t) \qquad p = \pm m\omega a \, \sinh(\omega t)$$
$$x \;=\; \pm a \, \sinh(\omega t) \qquad p = \pm m\omega a \, \cosh(\omega t)$$

The first case is for the particle with initial velocity towards the center of repulsion coming from $x = \pm\infty$ and stopping at $\pm a$ before rebounding. The energy $E = \frac{p^2}{2m} - \frac{kx^2}{2} = -\frac{ka^2}{2}$ is negative.

---

[9]M.G. Calkin, Lagrangian and Hamiltonian Mechanics, World Scientific Publishing Co. Pte. Ltd. (1996).

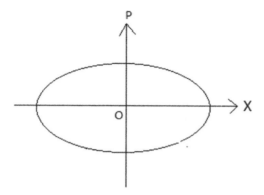

FIGURE 2.4: The ellipse.

The second case corresponds to $E = +\frac{ka^2}{2}$ which is positive. For $t = 0$ we have $x = 0$, $p = \pm m\omega a$, and the particle is with a finite velocity. Note that unlike ellipse, hyperbola is an open curve. See Figure 2.5. It is therefore clear why for the Cauchy problem, in the hyperbolic case, time appears as an open boundary. A summary of various features of $L(\phi) = 0$ and the conic represented by the equation $S(x, y) = 0$ is presented in Table 2.1 and Table 2.2 respectively.

TABLE 2.1: A summary of the various features of $L(\phi) = 0$.

System	Equation	Discriminant	Signature
PDE	$L(\phi) = 0$	$B^2 - AC$ (function of x, y only)	(i) $\quad -$ $E : (\frac{\partial^2}{\partial x^2} + \frac{\partial^2}{\partial y^2})\phi = 0$ (Laplace equation) imaginary characteristics Closed boundary (Dirichlet or Neumann) (ii) $\quad +$ $H : (\frac{\partial^2}{\partial x^2} - \frac{\partial^2}{\partial y^2})\phi = 0$ (Wave equation) real characteristics Open boundary (Cauchy) (iii) $\quad 0$ $P : \nabla^2\phi = \alpha^2\frac{\partial\phi}{\partial t}$ (Heat conduction equation) real characteristic Open boundary (Dirichlet or Neumann)

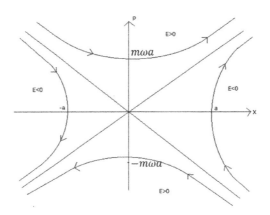

FIGURE 2.5: The hyperbola.

TABLE 2.2: A summary of the various features of the conic equation $S(x,y) = 0$ where $S(x,y) \equiv ax^2 + 2bxy + cy^2 + dx + ey + f$.

System	Equation	Discriminant		Signature
Conic	$S(x,y) = 0$	$b^2 - ac$	(i)	$-$
		(constant)		E: $\frac{x^2}{a^2} + \frac{y^2}{b^2} = 1$
				imaginary asymptotes
			(ii)	$+$
				H: $\frac{x^2}{a^2} - \frac{y^2}{b^2} = 1$
				real asymptotes
			(iii)	0
				P: $y^2 = 4ax$
				no asymptote

2.5   Adjoint and self-adjoint operators

We consider in a general way a second order linear operator L defined in terms of n independent variables

$$L = \sum_{i=1}^{n} \sum_{j=1}^{n} A_{ij} \frac{\partial^2}{\partial x_i \partial x_j} + \sum_{i=1}^{n} B_i \frac{\partial}{\partial x_i} + C \tag{2.73}$$

Note that the coefficient A_{ij} is symmetric and we assume that A_{ij}, B_i and C possess continuous second order derivatives to the independent variables $(x_1, x_2, ..., x_n)$.

Let ϕ and ψ be two arbitrary functions of the independent variables $(x_1, x_2, ..., x_n)$ which possess continuous derivatives. It then transpires that we can write

$$\psi A_{ij} \frac{\partial^2 \phi}{\partial x_i \partial x_j} = \frac{\partial}{\partial x_i}(\psi A_{ij} \frac{\partial \phi}{\partial x_j}) - \frac{\partial}{\partial x_j}[\phi \frac{\partial}{\partial x_i}(\psi A_{ij})] + \phi \frac{\partial^2}{\partial x_i \partial x_j}(\psi A_{ij}) \tag{2.74}$$

and

$$\psi B_i \frac{\partial \phi}{\partial x_i} = \frac{\partial}{\partial x_i}(B_i \psi \phi) - \phi \frac{\partial}{\partial x_i}(\psi B_i) \tag{2.75}$$

Taking now L between the functions ϕ and ψ gives

$$\psi L \phi = \phi [\sum_{i=1}^{n} \sum_{j=1}^{n} \frac{\partial^2}{\partial x_i \partial x_j}(\psi A_{ij}) - \sum_{i=1}^{n} \frac{\partial}{\partial x_i}(\psi B_i) + C\psi] +$$

$$+ \sum_{i=1}^{n} \frac{\partial}{\partial x_i}[\sum_{j=1}^{n}(\psi A_{ij} \frac{\partial \phi}{\partial x_j} - \sum_{j=1}^{n} \phi \frac{\partial}{\partial x_j}(A_{ij}\psi) + B_i \phi \psi] \tag{2.76}$$

where since A_{ij} is symmetric i.e. $A_{ij} = A_{ji}$ we have used

$$\sum_{i=1}^{n} \sum_{j=1}^{n} [\frac{\partial}{\partial x_j} \phi \frac{\partial}{\partial x_i}(A_{ij}\psi)] = \sum_{i=1}^{n} \sum_{j=1}^{n} [\frac{\partial}{\partial x_i} \phi \frac{\partial}{\partial x_j}(A_{ij}\psi)] \tag{2.77}$$

Calling

$$M\psi = \sum_{i=1}^{n} \sum_{j=1}^{n} \frac{\partial^2}{\partial x_i \partial x_j}(\psi A_{ij}) - \sum_{i=1}^{n} \frac{\partial}{\partial x_i}(\psi B_i) + C\psi \tag{2.78}$$

we can express $\psi L \phi$ as

$$\psi L \phi = \phi M \psi + \sum_{i=1}^{n} \frac{\partial}{\partial x_i}[\sum_{j=1}^{n} A_{ij}(\psi \frac{\partial \phi}{\partial x_j} - \phi \frac{\partial \psi}{\partial x_j}) + \phi\psi(B_i - \sum_{j=1}^{n} \frac{\partial A_{ij}}{\partial x_j})] \tag{2.79}$$

We thus have the Lagrange's identity

$$\psi L\phi - \phi M\psi = \sum_{i=1}^{n} \frac{\partial P_i}{\partial x_i} \tag{2.80}$$

where P_i stands for

$$P_i = \sum_{j=1}^{n} A_{ij}(\psi \frac{\partial \phi}{\partial x_j} - \phi \frac{\partial \psi}{\partial x_j}) + \phi\psi(B_i - \sum_{j=1}^{n} \frac{\partial A_{ij}}{\partial x_j}) \tag{2.81}$$

The operator M is called the adjoint to L. Note that the right side of (2.80) looks like a divergence. For instance, in two-dimensions, it looks like $(\frac{\partial P_x}{\partial x} + \frac{\partial P_y}{\partial y})$. However, the quantities P_x and P_y are functions of ϕ and ψ rather than the vector components. Still, the divergence form enables us to add and subtract to P_x and P_y respectively the quantities $-F_y$ and F_x, where F is an arbitrary function of (ϕ, ψ, x, y), without changing the right side of (2.80).

If L is a self-adjoint operator then $L = M$. In such a case we can express

$$M\phi = \sum_{i=1}^{n} \sum_{j=1}^{n} \frac{\partial^2}{\partial x_i \partial x_j}(\phi A_{ij}) - \sum_{i=1}^{n} \frac{\partial}{\partial x_i}(\phi B_i) + C\phi = L\phi + R \tag{2.82}$$

where R is

$$R = -2\sum_{j=1}^{n} \frac{\partial \phi}{\partial x_i}(B_i - \sum_{j=1}^{n} \frac{\partial A_{ij}}{\partial x_j}) - \phi \sum_{j=1}^{n} \frac{\partial}{\partial x_i}(B_i - \sum_{j=1}^{n} \frac{\partial A_{ij}}{\partial x_j}) \tag{2.83}$$

It enables us to conclude that the necessary and sufficient condition for the operator L to be self adjoint is that R vanishes which amounts to B_i being in the form

$$B_i = \sum_{j=1}^{n} \frac{\partial A_{ij}}{\partial x_j} \tag{2.84}$$

This means that L is expressible as

$$L = \sum_{i=1}^{n} \sum_{j=1}^{n} \frac{\partial}{\partial x_i}(A_{ij} \frac{\partial}{\partial x_j}) + C \tag{2.85}$$

Two-dimensional case

In the two-dimensional case corresponding to

$$L\phi = A\phi_{xx} + 2B\phi_{xy} + C\phi_{yy} + D\phi_x + E\phi_y + F\phi \tag{2.86}$$

the expression for its adjoint is

$$M\psi = (A\psi)_{xx} + 2(B\psi)_{xy} + (C\psi)_{yy} - (D\psi)_x - (E\psi)_y + F\psi \qquad (2.87)$$

where we have identified $A_{11} = A$, $A_{12} = B = A_{21}$, $A_{22} = C$, $B_1 = D$, $B_2 = E$, $C = F$. Also the quantities P_x and P_y are given by

$$P_x = A(\psi\phi_x - \phi\psi_x) + B(\psi\phi_y - \phi\psi_y) + (D - A_x - B_y)\phi\psi,$$
$$P_y = B(\psi\phi_x - \phi\psi_x) + C(\psi\phi_y - \phi\psi_y) + (E - B_x - C_y)\phi\psi \qquad (2.88)$$

Example 2.6

Given the L-operator to be $L = \frac{\partial^2}{\partial x^2} - p(x)\frac{\partial}{\partial x} + q(x)$, where $p(x)$ and $q(x)$ are twice continuously differentiable functions of x, find its adjoint M.

Comparing with the definition of the $L-$ operator given in the previous example, here $A = 1$, $D = -p$, $F = q$ and $B = C = E = 0$. Hence from the expression of M we have

$$M\psi = \psi_{xx} + (p\psi)_x + q\psi$$

Explicitly the M-operator reads

$$M = \frac{\partial^2}{\partial x^2} + p(x)\frac{\partial}{\partial x} + p_x + q(x)$$

2.6 Classification of PDE in terms of eigenvalues

If we focus on the second order terms of the L-operator (i.e. its principal part) we can identify

$$A = A_{11}, \quad B = A_{12} = A_{21}, \quad C = A_{22} \qquad (2.89)$$

Then the corresponding 2×2 matrix is

$$\begin{pmatrix} A & B \\ B & C \end{pmatrix}$$

whose eigenvalues λ_1, λ_2 are determined from the vanishing of its underlying determinant

$$\begin{vmatrix} A - \lambda & B \\ B & C - \lambda \end{vmatrix} = 0$$

These are

$$\lambda_1, \lambda_2 = \frac{A + C \pm \sqrt{(A-C)^2 + 4B^2}}{2} \tag{2.90}$$

which are clearly real because of the symmetric nature of the matrix. Their product is

$$\lambda_1 \lambda_2 = -(B^2 - AC) = -J \tag{2.91}$$

If $J > 0$ then the roots λ_1 and λ_2 are non-zero and are of opposite sign. The PDE is of hyperbolic type.

If $J < 0$ then the roots λ_1 and λ_2 are non-zero and are of same sign. The PDE is of elliptic type.

If $J = 0$ then one of the roots λ_1 and λ_2 is zero and the PDE is of parabolic type.

In general, for a nth order linear PDE we have to take into consideration its homogeneous part which corresponds to a real symmetric matrix. If we denote by P its number of positive roots and Z its number of zero roots, the classification of PDE goes as follows:

If $Z = 0$ and $P = 1$ or $Z = 0$ and $P = n - 1$, the PDE is of hyperbolic type.
If $Z > 0$ and $P = 0$ or $Z = 0$ and $P = n$, the PDE is of elliptic type.
If $Z = 0$, the PDE is of parabolic type.

Additionally if $Z > 0$ and $1 < P < n$, the PDE is of ultra-hyperbolic type. It is evident that for a PDE to be of ultra-hyperbolic type, it must involve at least four independent variables.

Example 2.7

The PDE

$$\phi_{xx} + \phi_{yy} - \phi_{zz} - \phi_{tt} - \phi_{ww} = 0$$

where (x, y, z, t, w) is a set of independent variables in $\Re^5$, is ultra-hyerbolic. This can be easily checked by examining the character of eigenvalues. The eigenvalue equation reads

$$(1 - \lambda)(1 - \lambda)(1 + \lambda)(1 + \lambda)(1 + \lambda) = 0$$

So there are 2 positive roots and no zero root. Since $0 < 2 < 5$, the PDE is ultra-hyperbolic.

2.7 Summary

The main purpose of this chapter was to show how the method of characteristics can be employed to reduce a second order linear PDE in two variables to a normal or canonical form. Indeed such a procedure provides an elegant way to solve the equation. We discussed in some detail the boundary and initial value problems and clarified the issues by working out a few typical problems. In this regard the feature of a well-posed problem was pointed out. We then provided some insights from classical mechanics by considering the oscillator problem under the influence of an elastic force and a repulsive one in the phase plane. In the former case the representative point moves in a closed curve while in the latter case an open curve is traced out. The roles of Dirichlet's (or Neumann) and Cauchy's conditions were discussed and analogies sought. We derived general expressions for the adjoint and self-adjoint operators. We also showed how a PDE can be classified in terms of the eigenvalues to the principal part of the second order operator.

Exercises

1. Consider the Tricomi equation given by

$$y\phi_{xx} + \phi_{yy} = 0, \quad y < 0$$

Bring it to a normal form.

2. Transform the telegraph equation

$$\phi_{tt} + a\phi_t + b\phi = \phi_{xx}$$

where a and b are constants to the normal form

$$\psi_{\xi\eta} + \frac{a^2 - 4b^2}{16c^2}\psi = 0$$

using the transformations $\xi(x,t) = x + ct$, $\eta(x,t) = x - ct$ and putting $\phi[x(\xi,\eta), t(\xi,\eta)] = \psi(\xi,\eta)e^{-\frac{at}{2}}$.

3. Find the general solution of the PDE

$$\phi_{xy} + \phi_y + 2x^2 y(\phi_x + \phi) = 0$$

4. Find the general solution of the PDE

$$e^{-2x}\phi_{xx} - e^{-2y}\phi_{yy} - e^{-2x}\phi_x + e^{-2y}\phi_y + 8e^y = 0$$

5. Solve the PDE

$$\phi_{xx} - 2\sin(x)\phi_{xy} - (3 + \cos^2(x))\phi_{yy} + \phi_x + (2 - \sin(x) - \cos(x))\phi_y = 0$$

subject to the conditions

$$\phi(x, \cos(x)) = 0, \quad \phi_y(x, \cos(x)) = e^{-\frac{x}{2}\cos(x)}$$

6. Solve the PDE

$$e^y\phi_{xy} - \phi_{yy} + \phi_y = 0$$

subject to the conditions

$$\phi_{xy}(x, 0) = 0, \quad \phi_y(x, 0) = -\sin(x)$$

7. Discuss the solution of the PDE

$$\phi_t + \phi\phi_x = -x, \quad t \geq 0$$

subject to the condition

$$\phi(x, 0) = f(x), \quad -\infty < x < \infty$$

8. Show that the L-operator operating as $L\phi(x, t) = \phi_{tt}(x, t) + \phi_{xxxx}(x, t)$ is self-adjoint.

9. If, as in non-relativistic quantum mechanics, the momentum operator has a representation $p = -i\partial_x$, then show that the Hamiltonian $H = x^3 p + px^3$ is formally self-adjoint.

Chapter 3

PDE: Elliptic form

As already noted in Chapter 2, the Poisson, Helmholtz and Laplace's equations are the three basic examples of an elliptic type of PDE. In practice they are used to model the time-invariant response of physical systems. Without a source term the Poisson equation reduces to the Laplace's equation whose solution, irrespective of its dimension, is referred to as a harmonic function.

The Laplace's equation

$$\nabla^2 \phi = 0 \quad \text{in} \quad \Omega \subset \Re^n \tag{3.1}$$

finds its natural appearence in problems of gravitation, electrostatics, magnetostatics and fluid dynamics. In two-dimensions, the function ϕ depends on the variables (x, y). Defining $z = x + iy$, the analyticity of a function $f(z)$ at a point $z = z_0$ means that it has a derivative at every point within a region encircling the point $z = z_0$. Let $f(x, y) = \psi(x, y) + i\phi(x, y)$. Then if the functions $\phi(x, y)$ and $\psi(x, y)$ satisfy the Cauchy-Riemann equations namely,

$$\frac{\partial \psi}{\partial x} = \frac{\partial \phi}{\partial y}, \quad \frac{\partial \psi}{\partial y} = -\frac{\partial \phi}{\partial x} \tag{3.2}$$

in a certain region where the partial derivatives are continuous, then $f(z)$ is analytic in that region. However, the converse may not be true i.e. if $f = \psi + i\phi$ satisfies Cauchy-Riemann equations then f may not be analytic. Take for example the function $f(x, y)$ defined by

$$f(x) = \begin{cases} \frac{x^3(1+i)-y^3(1-i)}{x^2+y^2} & \text{if } (x,y) \neq (0,0) \\ 0 & \text{if } (x,y) = (0,0) \end{cases}$$

then $f(x, y)$ satisfies the Cauchy-Riemann equations but f is not an analytic function.

Simple differentiations tell us that both ϕ and ψ obey Laplace's equations

$$\nabla^2\phi = \phi_{xx} + \phi_{yy} = 0, \quad \nabla^2\psi = \psi_{xx} + \psi_{yy} = 0 \tag{3.3}$$

We call the solutions ϕ and ψ of the Laplace's equation satisfying Cauchy-Riemann equations to be harmonic conjugates. Note that there is no dependence on time in ϕ which means that the Laplace's equation is concerned with a steady state picture. Because of symmetry reasons it is often advantageous to refer the Laplace's equation to polar coordinates. For instance, when we are dealing with an isolated point charge located at the origin, the potential produced by it at any point in space appears as inversely proportional to the radial distance from the origin whereas if the same problem is referred to in the Cartesian coordinates, a more complicated form, involving the square root of the sum of the squares of all the Cartesian coordinates for the point, emerges.

Poisson's equation also appears in problems of electrostatics, among other areas in theoretical physics. Let us consider the scalar potential χ at the point $\mathbf{x}$ in the form

$$\chi(\mathbf{x}) = \int \frac{\rho(\mathbf{x'})}{|\mathbf{x}\text{-}\mathbf{x'}|} d^3x' \tag{3.4}$$

for a continuous charge distribution characterized by the function ρ. Operating on both sides by ∇^2 and using the formula $\nabla^2(\frac{1}{|\mathbf{x}\text{-}\mathbf{x'}|}) = -4\pi\delta(\mathbf{x}\text{-}\mathbf{x'})$, Poisson's equation is easily seen to hold

$$\nabla^2\chi(\mathbf{x}) = -4\pi \int \rho(\mathbf{x'})\delta(\mathbf{x}\text{-}\mathbf{x'})d^3x' = -4\pi\rho(\mathbf{x}) \tag{3.5}$$

where ρ accounts for the inhomogeneity of the equation.

Conversely, Poisson's equation can be exploited to generate the type of charge distribution that could be relevant for a particular class of scalar potential. For instance if we look for the spherically symmetric Yukawa potential ϕ at the point $P(r)$

$$\chi(r) = \frac{e^{-\alpha r}}{r} \tag{3.6}$$

where r corresponds to the radial distance of P from the origin and α is a constant, then the charge distribution can be found out by substituting (3.6) into (3.5). To this end, we convert (3.5) to the polar form by referring to spherical coordinates

$$\frac{1}{r^2}\frac{d}{dr}(r^2\frac{d\chi}{dr}) = -4\pi\rho \tag{3.7}$$

and employ (3.6) in it. We then obtain for the continuous charge distribution the solution

$$\rho = \frac{\alpha^2}{4\pi}\frac{e^{-\alpha r}}{r} \tag{3.8}$$

Helmholtz's equation is given by

$$(\nabla^2 + k^2)\phi = 0 \tag{3.9}$$

where k is the wave number and ϕ represents the amplitude. Its solution[1] depends on the spatial boundary conditions and its applications are typically in the problems of wave mechanics and membranes. The paraxial approximation of the Helmholtz equation also plays an important role in quantum optics.

One of the objectives in this chapter will be to look at the boundary value problems that go with the elliptic equation. These problems are typically classified as the Dirichlet and Neumann problems. Consider a metallic plate having a closed curve C for its boundary which is kept at a constant temperature ϕ_0 i.e. $\phi = \phi_0$ on C. The temperature profile $\phi(x, y)$ obeys the two-dimensional Laplace's equation

$$\phi_{xx} + \phi_{yy} = 0 \tag{3.10}$$

Dirichlet problem associated with the two-dimensional Laplace's equation is the one when we are required to find the solution that takes prescribed values on the closed curve C. On the other hand, if the normal derivative of ϕ is prescribed on C, the problem is identified as a Neumann type. Of course in three-dimensions, the closed curve C is replaced by some closed surface.

We now turn to a discussion of the solvability criterion for the Laplace's equation. We begin with the method of separation of variables.

3.1 Solving through separation of variables

(a) Two dimensions: plane polar coordinates (r, θ)

In two-dimensional plane polar coordinates (r, θ), ∇^2 reads

$$\nabla^2 = \frac{\partial^2}{\partial r^2} + \frac{1}{r}\frac{\partial}{\partial r} + \frac{1}{r^2}\frac{\partial^2}{\partial\theta^2} \tag{3.11}$$

---

[1]The separable solution of Helmholtz's equation is discussed in Chapter 4 in connection with the separable spatial solution of the wave equation.

where in terms of Cartesian coordinates the variables r and θ read

$$r = \sqrt{x^2 + y^2}, \quad \theta = \tan^{-1}\left(\frac{y}{x}\right) \tag{3.12}$$

Note that $r \in (0, \infty)$ and $\theta \in (0, 2\pi]$.

We seek solutions of the two-dimensional Laplace's equation

$$\nabla^2 \phi(r, \theta) = 0 \tag{3.13}$$

in the variable-separated product form namely,

$$\phi(r, \theta) = F(r)G(\theta) \tag{3.14}$$

Substituting (3.14) in (3.13) and using (3.11) gives the ODE

$$\frac{1}{F}\left(r^2 \frac{d^2 F}{dr^2} + r \frac{dF}{dr}\right) = -\frac{1}{G}\frac{d^2 G}{d\theta^2} \tag{3.15}$$

Since the left side of (3.15) is a function of only r and the right side is a function of only θ, consistency requires that each side must assume a constant value which we take as n^2. (3.15) thus leads to the pair of decoupled equations

$$r^2 \frac{d^2 F}{dr^2} + r \frac{dF}{dr} - n^2 F = 0 \tag{3.16}$$

and

$$\frac{d^2 G}{d\theta^2} + n^2 G = 0 \tag{3.17}$$

(3.17) resembles the simple harmonic motion in classical mechanics. The solutions for F and G are straightforward to find and are given respectively by

$$F(r) = Ar^n + Br^{-n}, \quad n = 1, 2, \dots \tag{3.18}$$

and

$$G(\theta) = C \cos n\theta + D \sin n\theta, \quad n = 1, 2, \tag{3.19}$$

In the case when $n = 0$, the forms of the solutions are different

$$F(r) = A_0 \ln r + B_0, \quad n = 0 \tag{3.20}$$

and

$$G(\theta) = C_0, \quad n = 0 \tag{3.21}$$

Noting that if one added 2π to θ we would get back the same point we imposed

$$\phi(r, \theta) = \phi(r, \theta + 2\pi) \tag{3.22}$$

which implied that the integer values of n are relevant. We also omitted the θ-term in (3.21) because θ and $\theta + 2\pi$ represented the same point. Further, in (3.18) and (3.19) A, B, C, D and in (3.20) and (3.21) A_0, B_0, C_0 are constants.

Combining the above solutions we can put the separable form of the general solution of a two-dimensional Laplace's equation in the manner

$$\phi_0(r, \theta) = A_0 \ln r + B_0, \quad n = 0 \tag{3.23}$$

where C_0 has been absorbed in A_0 and B_0 and

$$\phi_n(r, \theta) = (Ar^n + Br^{-n})(C_n \cos n\theta + D_n \sin n\theta), \quad n = 1, 2, .. \tag{3.24}$$

The complete solution for $\phi(r, \theta)$ reads

$$\phi(r, \theta) = A_0 \ln r + \Sigma_n r^n (A_n \cos n\theta + B_n \sin n\theta) + \Sigma_n r^{-n}(C_n \cos n\theta + D_n \sin n\theta) + \Lambda_0 \tag{3.25}$$

where Λ_0 is a constant. Observe that (3.25) contains the following fundamental solution of the two dimensional Laplace's equation

$$\phi(r) = -\frac{1}{2\pi} \ln\left(\frac{1}{r}\right) \tag{3.26}$$

by fixing $A_0 = -\frac{1}{2\pi}$.

If we wish to avoid blowing up of the solution at $r = 0$ then ϕ would read from (3.23) and (3.24)

$$\phi(r, \theta) = \begin{cases} B_0, & n = 0 \\ \Sigma_n r^n (A_n \cos n\theta + B_n \sin n\theta), & n = 1, 2, ... \end{cases}$$

(b) Three dimensions: spherical polar coordinates (r, θ, ϕ)

In three-dimensional spherical polar coordinates (r, θ, ϕ), ∇^2 reads

$$\nabla^2 = \frac{1}{r^2} \frac{\partial}{\partial r}\left(r^2 \frac{\partial}{\partial r}\right) + \frac{1}{r^2 \sin \theta} \frac{\partial}{\partial \theta}\left(\sin \theta \frac{\partial}{\partial \theta}\right) + \frac{1}{r^2 \sin^2 \theta} \frac{\partial^2}{\partial \phi^2} \tag{3.27}$$

where in terms of Cartesian coordinates the variables r, θ and ϕ read

$$r = \sqrt{x^2 + y^2 + z^2}, \quad \theta = \cos^{-1}\left(\frac{z}{\sqrt{x^2 + y^2 + z^2}}\right), \quad \phi = \tan^{-1}\left(\frac{y}{x}\right) \tag{3.28}$$

Note that $r \in (0, \infty)$, $\theta \in [0, \pi)$ and $\phi \in [0, 2\pi)$.

We seek solutions of the three-dimensional Laplace's equation

$$\nabla^2 \psi(r, \theta, \phi) = 0 \tag{3.29}$$

in the variable-separated product form namely,

$$\psi(r, \theta, \phi) = R(r)f(\theta, \phi) \tag{3.30}$$

Substituting (3.30) in (3.29) and using (3.27) gives

$$\frac{1}{R}\frac{d}{dr}(r^2\frac{dR}{dr}) = -\frac{1}{f}[\frac{1}{\sin\theta}\frac{\partial}{\partial\theta}(\sin\theta\frac{\partial f}{\partial\theta}) + \frac{1}{\sin^2\theta}\frac{\partial^2 f}{\partial\phi^2}] = -p(\text{say}) \tag{3.31}$$

where p is a constant because the left side of the first equality deals with a function of r only while the second equality depends on a function of θ and ϕ only and these are consistent if each is equal to a constant.

The R-equation is given by

$$\frac{1}{R}\frac{d}{dr}(r^2\frac{dR}{dr}) + p = 0 \tag{3.32}$$

which amounts to the form

$$\frac{d^2 R}{ds^2} + \frac{dR}{ds} - l(l+1)R = 0 \tag{3.33}$$

where we have defined $p = -l(l+1)$ and set $s = \ln r$.

The second order ODE (3.33) has the general solution

$$R(r) = Ar^l + Br^{-(l+1)} \tag{3.34}$$

where A and B are arbitrary constants.

We now separate $f(r, \theta)$ in the product form

$$f(r, \theta) = g(\theta)h(\phi) \tag{3.35}$$

Substituting in (3.31) one obtains from the second equation

$$\frac{\sin\theta}{g}\frac{d}{d\theta}(\sin\theta\frac{dg}{d\theta}) + l(l+1)\sin^2\theta = -\frac{1}{h}\frac{d^2 h}{d\phi^2} = \nu^2(\text{say}) \tag{3.36}$$

where we have applied a similar reasoning as noted below (3.31) that since we have two equations, one in θ and another in ϕ, the two can be compromised if each is equal to a constant which we have set as ν^2.

As a result of (3.36), the differential equations for h and g assume respectively the forms

$$\frac{d^2 h}{d\phi^2} = -h\nu^2 \tag{3.37}$$

and

$$\frac{d}{d\mu}[(1-\mu^2)\frac{dg}{d\mu}] + [l(l+1) - \frac{\nu^2}{1-\mu^2}]g == 0 \tag{3.38}$$

where we have put $\mu = \cos\theta$.

The general solution of (3.37) is given by the periodic form

$$h(\phi) = E\cos(\nu\phi) + F\sin(\nu\phi) \tag{3.39}$$

where E and F are arbitrary constants.

On the other hand, equation (3.38) can be immediately identified as an associated Legendre equation whose general solution is well known to be

$$g(\mu) = CP_l^\nu(\mu) + DQ_l^\nu(\mu) \tag{3.40}$$

where $P_l^\nu(\mu)$ and $Q_l^\nu(\mu)$ are the associated Legendre functions which are expressible in terms of hypergeometric functions. In (3.40), C and D are two arbitrary constants.

Since we are often interested in a physically meaningful finite and single-valued solution[2] in the ranges $0 \le \phi < 2\pi$ and $0 \le \theta < \pi$ we restrict $\nu = m$, $m = 0, 1, 2, ...$ In other words, ν must be taken as a non-negative integer. Further, because of logarithmic singularities at $\mu = \pm 1$ in $Q_l^\nu(\mu)$, we set $D = 0$, For the series representation of $P_l^m(\mu)$ to be convergent the restriction on l is that it must be a positive integer or zero. In such a case the infinite series for $P_l^m(\mu)$ terminates and is given by the relation

$$P_l^m(\mu) = (1-\mu^2)^{\frac{m}{2}}\frac{d^m}{d\mu^m}P_l(\mu), \quad l = 0, 1, 2, ... \tag{3.41}$$

where $P_l(\mu)$ obeys the Rodrigues formula

$$P_l(\mu) = \frac{1}{2^l l!}\frac{d^l}{d\mu^l}(\mu^2 - 1)^l \tag{3.42}$$

The general solution of the Laplace's equation in spherical polar coordinates that remains finite and single-valued in a region that includes the origin is

$$\psi(r,\theta,\phi) = \sum_{l=0}^{\infty}r^l\sum_{m=0}^{l}(A_{lm}\cos(m\phi) + B_{lm}\sin(m\phi))P_l^m(\mu) = \sum_{l=0}^{\infty}r^l Y_l(\theta,\phi), \quad l = 0, 1, 2, ...$$

where Y_l, the surface spherical harmonics of degree n, and stands for the angular part

---

[2]The physical requirement of single valuedness concerning $h(\phi)$ is $h(\phi + 2\pi) = h(\phi)$ resulting in ν to be zero or integer.

$$Y_l(\theta, \phi) = \sum_{m=0}^{l} (A_{lm} \cos(m\phi) + B_{lm} \sin(m\phi)) P_l^m(\mu)$$

The spherical harmonics are orthogonal and normalized

$$\int_0^{2\pi} \int_0^{\pi} d\phi d\theta \sin\theta Y_l^m(\theta, \phi)^* Y_{l'}^{m'}(\theta, \phi) = \delta_{ll'} \delta_{mm'}$$

where we took the form $Y_l^m(\theta, \phi) = N e^{im\phi} P_l^m(\cos\theta)$, N is a normalization constant to be chosen so as to have the right side of the integral equal unity when $l = l', m = m'$.

On the other hand, the general solution of the Laplace's equation in spherical polar coordinates that remains finite and single-valued in a region that includes the point at infinity is

$$\phi(r, \phi) = \sum_{l=0}^{\infty} r^{-l-1} Y_l(\theta, \phi), \quad l = 0, 1, 2, ...$$

The fundamental solution of Laplace's equation in three-dimensions corresponds to

$$\phi(r) = -\frac{1}{4\pi} \frac{1}{r}, \quad r \neq 0 \tag{3.43}$$

while for the n-dimensional case, the fundamental solution is

$$\phi(r) = \frac{1}{n(n-2)B(n)} \frac{1}{r^{n-2}}, \quad r \neq 0, \quad n \geq 3 \tag{3.44}$$

where $B(n)$ stands for the volume of a unit ball in $\Re^n$.

(c) Cylindrical polar coordinates (r, θ, z)

In cylindrical polar coordinates (r, θ, z), ∇^2 reads

$$\nabla^2 = \frac{\partial^2}{\partial r^2} + \frac{1}{r}\frac{\partial}{\partial r} + \frac{1}{r^2}\frac{\partial^2}{\partial \theta^2} + \frac{\partial^2}{\partial z^2} \tag{3.45}$$

where in terms of the Cartesian coordinates the variables r, θ and z are given by

$$r = \sqrt{x^2 + y^2}, \quad \theta = \tan^{-1}(\frac{y}{x}), \quad z = z \tag{3.46}$$

Note that $r \in [0, \infty)$, $\theta \in [0, 2\pi)$ and $z \in (\infty, \infty)$.

We seek solutions of the three-dimensional Laplace's equation

$$\nabla^2 \phi(r, \theta, z) = 0 \tag{3.47}$$

in the variable-separated product form namely,

$$\phi(r, \theta) = R(r) f(\theta) h(z) \tag{3.48}$$

Substituting (3.48) in (3.47) and using (3.45) gives the form

$$\frac{1}{h}\frac{d^2h}{dz^2} = -\frac{1}{R}\left(\frac{d^2R}{dr^2} + \frac{1}{r}\frac{dR}{dr}\right) - \frac{1}{fr^2}\frac{d^2f}{d\theta^2} \tag{3.49}$$

Since the left side is a function of z only and the right side is a function of r and θ consistency requires that each side is equal to a constant which we set as p^2. Then we get the following pair of equations

$$\frac{d^2h}{dz^2} = p^2h \tag{3.50}$$

and

$$\frac{r^2}{R}\frac{d^2R}{dr^2} + \frac{r}{R}\frac{dR}{dr} + p^2r^2 = -\frac{1}{f}\frac{d^2f}{d\theta^2} \tag{3.51}$$

The solution of (3.50) is known to be in the form

$$h(z) = \alpha e^{pz} + \beta e^{-pz} \tag{3.52}$$

where α and β are arbitrary constants.

In (3.51) the variables are already separated and so we set each side as equal to the constant ν^2. This gives

$$\frac{d^2f}{d\theta^2} + \nu^2 f = 0 \tag{3.53}$$

and

$$r^2\frac{d^2R}{dr^2} + r\frac{dR}{dr} + (p^2r^2 - \nu^2)R = 0 \tag{3.54}$$

Equation (3.53) has the simple harmonic form whose general solution reads

$$f(\theta) = \gamma\cos(\nu\theta) + \delta\sin(\nu\theta) \tag{3.55}$$

where γ and δ are arbitrary constants. If one requires a unique θ then $f(\theta+2\pi) = f(\theta)$ results in ν to be zero or an integer.

The differential equation (3.54) can be easily converted to the Bessel form by applying a transformation

$$y = pr \tag{3.56}$$

which gives

$$y^2\frac{d^2R}{dy^2} + y\frac{dR}{dy} + (y^2 - \nu^2)R = 0 \tag{3.57}$$

Its general solution for all values of ν is

$$R(r) = \lambda J_\nu(pr) + \mu Y_\nu(pr) \tag{3.58}$$

where J_ν and Y_ν are respectively Bessel and Neumann functions of order ν.

It is useful to note that had we chosen a negative separation constant $-p^2$ we would have gotten a modified Bessel equation

$$y^2 \frac{d^2 R}{dy^2} + y \frac{dR}{dy} - (y^2 + \nu^2)R = 0 \qquad (3.59)$$

whose solutions are the modified Bessel functions I_ν and K_ν

$$R(r) == \lambda' I_\nu(pr) + \mu' K_\nu(pr) \qquad (3.60)$$

Both K_ν and Y_ν are divergent at $r = 0$ and so are to be excluded if we are interested in a solution around $r = 0$. On the other hand, because J_ν and I_ν diverge at $r \to \infty$ they are to be disregarded for an exterior solution.

3.2   Harmonic functions

Simple algebraic functions like $\phi = x^3 - 3xy^2$ and $\phi = 3x^2y - y^3$ or transcendental functions such as $\phi = e^x \cos(y)$ and $\phi = e^x \sin(y)$ which satisfy Laplace's equation are examples of harmonic functions in two variables while a unit point charge at origin described by the function $\frac{1}{r}$ in a three-dimensional region $(r \neq 0)$ and a line of unit charge density on the whole of z-axis in terms of the function $-\ln(r^2 - z^2)$, $r \neq z$, are examples of harmonic functions in three variables.

We shall now prove Gauss' mean value theorem for a harmonic function.

Gauss' mean value theorem

Let ϕ be a harmonic function in a three-dimensional region $\Re^3$ and P be a point inside it. We imagine a sphere S of radius r with centre at P which is immersed wholly inside $\Re^3$. Then ϕ at P is given by the integral

$$\phi(P) = \frac{1}{4\pi r^2} \int \int_S \phi(Q) ds \qquad (3.61)$$

where ds is the surface element and Q is a point on S. The above result is called Gauss' mean value theorem.

Proof :   The spherical mean of ϕ over the surface of the sphere S is

$$\bar{\phi}(r) = \frac{1}{4\pi r^2} \int \int_S \phi(Q) ds \qquad (3.62)$$

If the Cartesian coordinates of P are given by (x, y, z) then those of Q are evidently (ξ, η, ζ) where

$$\xi = x + lr, \quad \eta = y + mr, \quad \zeta = z + nr \tag{3.63}$$

with the quantities l, m, n, in terms of spherical polar angles θ and ϕ, stand for

$$l = \sin\theta\cos\phi, \quad m = \sin\theta\sin\phi, \quad n = \cos\theta \tag{3.64}$$

Noting that the surface element ds is $r^2 \sin\theta d\theta d\phi$, $\bar{\phi}(r)$ reads

$$\bar{\phi}(r) = \frac{1}{4\pi} \int_0^{2\pi} \int_0^{\pi} \phi(x + lr, y + mr, z + nr) \sin\theta d\theta d\phi \tag{3.65}$$

By differentiating with respect to r one finds

$$\frac{d\bar{\phi}(r)}{dr} = \frac{1}{4\pi} \int_0^{2\pi} \int_0^{\pi} (l\frac{\partial\phi}{\partial\xi} + m\frac{\partial\phi}{\partial\eta} + n\frac{\partial\phi}{\partial\zeta}) \sin\theta d\theta d\phi \tag{3.66}$$

which can be recast as

$$\frac{d\bar{\phi}(r)}{dr} = \frac{1}{4\pi r^2} \int_0^{2\pi} \int_0^{\pi} (l\frac{\partial\phi}{\partial\xi} + m\frac{\partial\phi}{\partial\eta} + n\frac{\partial\phi}{\partial\zeta}) ds \tag{3.67}$$

The above representation prompts us to make use of the divergence theorem to convert the surface integral into the following volume integral

$$\frac{d\bar{\phi}(r)}{dr} = \frac{1}{4\pi r^2} \int\int\int \nabla^2 \phi dv \tag{3.68}$$

where dv is the volume element.

The harmonic property of ϕ makes the right side vanish and we have the simple result

$$\frac{d\bar{\phi}(r)}{dr} = 0 \tag{3.69}$$

In other words, $\bar{\phi}(r)$ is a constant function independent of r. Hence it is possible to interpret

$$
\begin{aligned}
\bar{\phi}(r) &= \bar{\phi}(0) \\
&= \lim_{r\to 0} \bar{\phi}(r) \\
&= \lim_{r\to 0} \frac{1}{4\pi} \int_0^{2\pi} \int_0^{\pi} \phi(x + lr, y + mr, z + nr) \sin\theta d\theta d\phi \\
&= \frac{1}{4\pi} \int_0^{2\pi} \int_0^{\pi} \lim_{r\to 0} \phi(x + lr, y + mr, z + nr) \sin\theta d\theta d\phi \\
&= \frac{1}{4\pi} \phi(x, y, z) \int_0^{2\pi} \int_0^{\pi} \sin\theta d\theta d\phi \\
&= \phi(x, y, z) \\
&= \phi(P) \tag{3.70}
\end{aligned}
$$

where use has been made of the double-integral having the value of 4π. From (3.62) we therefore conclude that

$$\phi(P) = \frac{1}{4\pi r^2} \int \int_S \phi(Q)ds \qquad (3.71)$$

By a similar reasoning the result in two-dimensions can be proved.

3.3   Maximum-minimum principle for Poisson's and Laplace's equations

Let us consider a two-dimensional Poisson equation

$$\nabla^2\phi = -4\pi\rho(x, y) \qquad (3.72)$$

in a given bounded region Ω having the boundary $\partial\Omega$. We assume ϕ to be continuous in Ω as well as on $\partial\Omega$. The maximum principle states that if $\rho < 0$ in Ω and ϕ is a solution of the Poisson equation then its maximum value is attained on $\partial\Omega$. On the other hand, if $\rho > 0$ in Ω and ϕ is a solution of the Poisson equation then its minimum value is attained on $\partial\Omega$. We give a simple proof of these assertions.

Proof: First let $\rho < 0$. The continuity of ϕ implies that its maximum ought to exist anywhere in $\Omega + \partial\Omega$. If it occurs at a point $(\overline{x}, \overline{y})$ inside Ω then evidently $(\phi_x)_{(\overline{x},\overline{y})} = 0 = (\phi_y)_{(\overline{x},\overline{y})}$ along with $(\phi_{xx})_{(\overline{x},\overline{y})} < 0$ and $(\phi_{yy})_{(\overline{x},\overline{y})} < 0$ which by (3.72) contradicts the assertion that $\rho < 0$. Hence the maximum of ϕ is attained on the boundary.

Similarly we can establish that the minimum of $\phi(x, y)$ is attained on the boundary for the opposite case $\rho > 0$. Note that the latter amounts to replacing ϕ by $-\phi$ in (3.72) . Since $\rho > 0$ is equivalent to $-\rho < 0$, the preceding arguments lead to $-\phi$ assuming its maximum value on $\partial\Omega$. In other words, the minimum of $\phi(x, y)$ is attained on the boundary $\partial\Omega$.

We now turn to Laplace's equation which corresponds to $\rho = 0$. Let us suppose that the maximum M of ϕ occurs on the boundary $\partial\Omega$. Define a function $\psi = \phi + \epsilon(x^2 + y^2)$, $\epsilon > 0$ which implies $\psi \geq \phi$. The function is often referred to as the helper function.

Operating on both sides of ψ by ∇^2 we find $\nabla^2\psi = 4\epsilon > 0$ because of $\nabla^2(x^2 + y^2) = 4$. Thus ψ satisfies the Poisson's equation in Ω and so the maximum-minimum principle for ψ holds. We therefore conclude that the maximum of ψ has to occur on the boundary $\partial\Omega : \psi \leq M + \epsilon R^2$ where $x^2 + y^2 = R^2$. With $\psi \geq \phi$ it thus follows that $\phi \leq M + \epsilon R^2 \Rightarrow \phi \leq M$ in $\partial\Omega$ on letting $\epsilon \to 0$. In other words, if $\phi \leq M$ on $\partial\Omega$ then $\phi \leq M$ in Ω too.

Now if ϕ is a solution of Laplace's equation, then $- \phi$ is a solution too and so the minimum principle follows trivially.

As a corollary, we can comment on the stability of the solution. Let ϕ be a solution of the Poisson's equation in $\Omega + \partial\Omega$. Consider ϕ' to be another solution that results on perturbing the Dirichlet boundary condition on $\partial\Omega$ by a small amount. Evidently $|\phi - \phi'|$ is small too on $\partial\Omega \Rightarrow |\phi - \phi'|_{\partial\Omega} < \epsilon \Rightarrow |\phi - \phi'| < \epsilon$ throughout Ω and the stability of the solution is implied. (Note that $(\phi - \phi')$ satisfies the Laplace equation in $\partial\Omega$ and so the maximum of $|\phi - \phi'|$ is attained on $\partial\Omega$ where it is less than ϵ.)

3.4 Existence and uniqueness of solutions

Let us focus on the Poisson's equation $\nabla^2\phi = -4\pi$ (we have put $\rho = 1$) and assume the existence of its solution in the region Ω which has the boundary $\partial\Omega$. Our aim is to establish the following theorem:

Theorem

A unique solution exists within Ω which on its closed boundary $\partial\Omega$ satisfies either the condition of Dirichlet (or a mixed one) or Neumann (up to a constant). However, should $\partial\Omega$ be open, Dirichlet or Neumann boundary condition is insufficient to produce a unique solution. The problem is overdetermined if we apply Cauchy boundary conditions on the closed surface.

Proof :   Given that the Poisson's equation is defined for the region Ω having the boundary $\partial\Omega$, let, if possible, two solutions exist which are specified by ϕ_1 and ϕ_2. Then their difference $\Phi \equiv \phi_1 - \phi_2$ satisfies Laplace's equation $\nabla^2\Phi = 0$ on $\bar{\Omega}$, the closure of Ω, along with $\Phi = 0$ on $\partial\Omega$ for the Dirichlet boundary condition and $\frac{\partial\Phi}{\partial n} = 0$ on $\partial\Omega$ for the Neumann boundary condition. For the mixed boundary condition on $\Omega = \Omega_1 + \Omega_2$ with $\partial\Omega_1$ and $\partial\Omega_2$ corresponding to two open surfaces that constitute the whole of $\partial\Omega$ i.e. $\partial\Omega = \partial\Omega_1 + \partial\Omega_2$ such that $\Phi = 0$ on $\partial\Omega_1$ and $\frac{\partial\Phi}{\partial n} = 0$ on $\partial\Omega_2$, it follows that in these cases $\Phi\frac{\partial\Phi}{\partial n} = 0$ on $\partial\Omega$. Taking cue from the latter we observe that

$$
\begin{aligned}
0 &= \oint_{\partial\Omega} \Phi\frac{\partial\Phi}{\partial n}\, dS \\
&= \oint_{\partial\Omega} \Phi\nabla\Phi.\hat{n}\, dS \\
&= \int_{\Omega} \nabla.(\Phi\nabla\Phi)\, dV \qquad \text{(using Gauss' theorem)} \\
&= \int_{\Omega} (\Phi\nabla^2\Phi + |\nabla\Phi|^2)\, dV \\
&= \int_{\Omega} |\nabla\Phi|^2\, dV \qquad (\because \nabla^2\Phi = 0) \qquad (3.73)
\end{aligned}
$$

Consistency of both sides requires that $|\nabla\Phi|^2 = 0 \Rightarrow \nabla\Phi = 0$ on Ω. This means that $\Phi =$ constant on Ω. For the Dirichlet case, this constant is zero. The constant is zero too in the mixed case because in the Ω_1 portion of Ω, $\Phi = 0$. However, for the Neumann case any value of the constant is feasible because whatever the value of the constant, $\frac{\partial \Phi}{\partial n} = 0$.

Consider employing a Dirichlet condition on the open surface $\partial \Omega_1$

$$\phi|_{\partial \Omega_1} = f(x, y, z) \qquad (3.74)$$

and suppose for $\partial \Omega_2$ the following feature holds

$$\phi|_{\partial \Omega_2} = g_1(x, y, z) \qquad (3.75)$$

The above conditions on $\partial \Omega_1$ and $\partial \Omega_2$ thus provide a Dirichlet condition on the closed $\partial \Omega$. The solution pertaining to this Dirichlet condition is evidently unique in Ω as we learnt from the previous proposition. Let us specify this unique solution by $\phi = \phi_1(x, y, z)$.

We now modify the condition on $\partial \Omega_2$ to read

$$\phi|_{\partial \Omega_2} = g_2(x, y, z) \qquad (3.76)$$

Applying a similar reasoning on this along with the condition on $\partial \Omega_1$ we have another Dirichlet condition on the closed surface $\partial \Omega$ which is also unique in Ω. Let us call this solution $\phi = \phi_2(x, y, z)$. Note that ϕ_1 and ϕ_2 are different since they differ on $\partial \Omega_2$. We therefore observe that both ϕ_1 and ϕ_2 satisfy the same PDE supported by a Dirichlet condition on an open $\partial \Omega_1$. We are thus led to the non-uniqueness of the solution. Carrying on the argument we would run into an infinity of solutions.

Regarding the Cauchy boundary conditions on the closed surface Ω, let us specify them by

$$\phi|_{\partial \Omega} = f_1(x, y, z), \quad \frac{\partial \phi}{\partial n}|_{\partial \Omega} = f_2(x, y, z) \qquad (3.77)$$

If we consider the first condition only then we have a Dirichlet problem for the given PDE and from what we discussed above, a unique solution for it exists within Ω. Let us call this solution $\phi_1(x, y, z)$. This solution allows us to determine the normal derivative $\frac{\partial \phi_1}{\partial n}$ on Ω. If it coincides with the given function $f_2(x, y, z)$ then we have a solution for the Cauchy boundary conditions but at the cost of rendering one of the two prescribed conditions to be unnecessary. The problem is therefore overdetermined. If the coincidence does not happen then we do not have any solution.

3.5   Normally directed distribution of doublets

Consider a circle of radius ϵ with its centre at the fixed point $P(x, y)$ and having a boundary C lying entirely inside a two-dimensional region Ω whose boundary is $\partial \Omega$. Then in the sub-region Ω' formed by omitting C, the fundamental solution $-\frac{1}{2\pi} \ln \frac{1}{r}$ of the two-dimensional Laplace's equation holds

$$\left(\frac{\partial^2}{\partial \xi^2} + \frac{\partial^2}{\partial \eta^2}\right) \ln \frac{1}{r} = 0, \quad (\xi, \eta) \in \Omega' \tag{3.78}$$

where $r^2 = (x - \xi)^2 + (y - \eta)^2$ and (ξ, η) are the coordinates of the variable integration point.

We now employ the following Green's second identity

$$\int\int_\Omega (u\nabla^2 v - v\nabla^2 u) d\Omega = \int_{\partial\Omega} (u\frac{\partial v}{\partial n} - v\frac{\partial u}{\partial n}) dl \tag{3.79}$$

where the functions u and v have continuous second derivatives in $\bar{\Omega}$. In Ω', which is a multiply-connected region because of the non-intersecting closed boundaries $\partial\Omega'$ and C, if we identify $u = \phi$ and $v = \ln \frac{1}{r}$, then we obtain the formula

$$-\int\int_{\Omega'} (\ln \frac{1}{r})\nabla^2\phi d\Omega' = \int_{\partial\Omega'} [\phi\frac{\partial}{\partial n}(\ln \frac{1}{r}) - (\ln \frac{1}{r})\frac{\partial\phi}{\partial n}] dl' + \int_C [\phi\frac{\partial}{\partial n}(\ln \frac{1}{r}) - (\ln \frac{1}{r})\frac{\partial\phi}{\partial n}] dl \tag{3.80}$$

where we have used (3.78).

We make a couple of observations:

(i) Since ϕ has continuous derivatives in Ω, it gives the bound $|\frac{\partial\phi}{\partial n}| \leq \Lambda$, where Λ is a constant in Ω. Accordingly,

$$\int_C |-(\ln \frac{1}{r})\frac{\partial\phi}{\partial n}| dl \leq \ln \epsilon \int_C |\frac{\partial\phi}{\partial n}| dl = 2\pi\Lambda\epsilon \ln \epsilon \to 0 \quad \text{as} \quad \epsilon \to 0 \tag{3.81}$$

(ii) Since on C, $\frac{\partial}{\partial n} = -\frac{\partial}{\partial r}$, we have on taking the derivative

$$\int_C \phi\frac{\partial}{\partial n}(\ln \frac{1}{r}) dl = \frac{1}{\epsilon}\int_0^{2\pi} \phi(x+\epsilon\cos\theta, y+\epsilon\sin\theta)\epsilon d\theta = \int_0^{2\pi} \phi(x+\epsilon\cos\theta, y+\epsilon\sin\theta) d\theta \tag{3.82}$$

With ϕ being a continuous function, the integrand above is a continuous function too. So taking the limit through the sign of integration we can write

$$\lim_{\epsilon\to 0} \int_C \phi\frac{\partial}{\partial n}(\ln \frac{1}{r}) dl = \int_0^{2\pi} \lim_{\epsilon\to 0} \phi(x + \epsilon\cos\theta, y + \epsilon\sin\theta) d\theta \tag{3.83}$$

implying

$$\lim_{\epsilon\to 0} \int_C \phi\frac{\partial}{\partial n}(\ln \frac{1}{r}) dl = 2\pi\phi(x, y) \tag{3.84}$$

Inserting (3.81) and (3.84) in (3.80) we have the result

$$\phi(x, y) = -\frac{1}{2\pi}\int\int_\Omega (\ln \frac{1}{r})\nabla^2\phi d\Omega - \frac{1}{2\pi}\int_{\partial\Omega} [\phi\frac{\partial}{\partial n}(\ln \frac{1}{r}) - (\ln \frac{1}{r})\frac{\partial\phi}{\partial n}] dl \tag{3.85}$$

in the limit $\epsilon \to 0$.

The interpretation of the right side of (3.85) goes as follows:

The first and third terms represent respectively the potential of surface distribution and line distribution of matter of density $-\frac{1}{2\pi}\nabla^2\phi$ and $\frac{1}{2\pi}\frac{\partial\phi}{\partial n}$, one is distributed over the surface Ω and the other along $\partial\Omega$. The second term is due to a normally directed distribution of doublets of strength per unit length $-\frac{\phi}{2\pi}$ along $\partial\Omega$.

For the case when $\phi(x, y)$ is a harmonic function then the first term in the right side of (3.85) drops out and we have the reduced expression

$$\phi(x, y) = -\frac{1}{2\pi}\int_{\partial\Omega}[\phi\frac{\partial}{\partial n}(\ln\frac{1}{r}) - (\ln\frac{1}{r})\frac{\partial\phi}{\partial n}]dl \qquad (3.86)$$

It implies a superposition of potentials of normally directed distribution of doublets and a distribution of matter.

For the three-dimensional case the procedure is similar. Here we replace the circle by a sphere S of radius ϵ which lies entirely inside a three-dimensional region Ω having a boundary $\partial\Omega$. The sphere has its centre at the fixed point $P(x, y, z)$ and has a boundary S. Then in the sub-region Ω' formed by omitting S, the fundamental solution $-\frac{1}{4\pi}\frac{1}{r}$ of the three-dimensional Laplace's equation holds

$$(\frac{\partial^2}{\partial\xi^2} + \frac{\partial^2}{\partial\eta^2} + \frac{\partial^2}{\partial\zeta^2})\frac{1}{r} = 0, \quad (\xi, \eta, \zeta) \in \Omega' \qquad (3.87)$$

where $r^2 = (x - \xi)^2 + (y - \eta)^2 + (z - \zeta)^2$ and (ξ, η, ζ) are the coordinates of the variable integration point.

We now use Green's second identity (3.79) for the region Ω' which has non-interacting closed boundaries $\partial\Omega'$ and S:

$$\int\int\int_{\Omega'}(u\nabla^2 v - v\nabla^2 u)d\Omega' = \int\int_{\partial\Omega'}(u\frac{\partial v}{\partial n} - v\frac{\partial u}{\partial n})dS' + \int\int_S(u\frac{\partial v}{\partial n} - v\frac{\partial u}{\partial n})dS \qquad (3.88)$$

where we have identified $u = \phi$ and $v = \frac{1}{r}$. Because of (3.87) we are led to

$$-\int\int\int_{\Omega'}(\frac{1}{r})\nabla^2\phi d\Omega' = \int\int_{\partial\Omega'}[\phi\frac{\partial}{\partial n}(\frac{1}{r}) - (\frac{1}{r})\frac{\partial\phi}{\partial n}]dS'$$
$$+ \int\int_S[\phi\frac{\partial}{\partial n}(\ln\frac{1}{r}) - (\ln\frac{1}{r})\frac{\partial\phi}{\partial n}]dS \qquad (3.89)$$

Since ϕ has continuous derivatives in Ω, employing similar arguments as in the two-dimensional case, we can see that the last term in the right side of (3.89) goes to zero as $\epsilon \to 0$. Similarly on S, the derivative $\frac{\partial}{\partial n} = -\frac{\partial}{\partial R}$. The arguments of ϕ being $(x + \epsilon\sin\theta\cos\phi, y + \epsilon\sin\theta\sin\phi, z + \epsilon\cos\theta)$ we have on taking the limit $\epsilon \to 0$

$$\lim_{\epsilon\to 0}\int\int_S\phi\frac{\partial}{\partial n}(\frac{1}{r}) = 4\pi\phi(x, y, z) \qquad (3.90)$$

Thus, in the limit $\epsilon \to 0$, we have from (3.89), the result

$$\phi(x,y,z) = -\frac{1}{4\pi}\int\int\int_\Omega (\frac{1}{r})\nabla^2\phi d\Omega - \frac{1}{4\pi}\int\int_{\partial\Omega}[\phi\frac{\partial}{\partial n}(\frac{1}{r}) - (\frac{1}{r})\frac{\partial\phi}{\partial n}]dS \quad (3.91)$$

The terms in the right hand side of (3.91) can be interpreted as follows:

The first and third terms represent the potential of volume distribution and surface distribution of matter of density $-\frac{1}{4\pi}\nabla^2\phi$ and $\frac{1}{4\pi}\frac{\partial\phi}{\partial n}$ respectively, one is distributed over the volume Ω and the other over $\partial\Omega$. The second term is due to the potential of a double layer of density $-\frac{\phi}{4\pi}$ distributed over $\partial\Omega$.

Finally, should $\phi(x,y)$ be a harmonic function the first term in the right side of (3.91) drops out and we have the formula

$$\phi(x,y,z) = -\frac{1}{4\pi}\int\int_{\partial\Omega}[\phi\frac{\partial}{\partial n}(\frac{1}{r}) - (\frac{1}{r})\frac{\partial\phi}{\partial n}]dS \quad (3.92)$$

which is known as the Green's equivalent layer theorem. It gives the solution of the boundary value problem for Laplace's equation in three-dimensions when the values of ϕ and $\frac{\partial\phi}{\partial n}$ are given on the surface $\partial\Omega$.

3.6   Generating Green's function for Laplacian operator

In two-dimensions we already have for $\phi(x,y)$ the result (3.85). Let us take in addition a two-dimensional harmonic function $\psi(\xi,\eta)$ in Ω which is twice continuously differentiable in $\bar\Omega$ ($\equiv \Omega + \partial\Omega$). We then obtain from Green's second identity (3.79) the relation

$$-\frac{1}{2\pi}\int\int_\Omega \psi\nabla^2\phi d\Omega = -\frac{1}{2\pi}\int_{\partial\Omega}(\phi\frac{\partial\psi}{\partial n} - \psi\frac{\partial\phi}{\partial n})dl \quad (3.93)$$

on putting $u = \phi$ and $v = \psi$.

Subtracting it from (3.85) we derive another representation for $\phi(x,y)$

$$\phi(x,y) = -\frac{1}{2\pi}\int\int_\Omega (\psi - \ln r)\nabla^2\phi d\Omega - \frac{1}{2\pi}\int_{\partial\Omega}[(\phi\frac{\partial}{\partial n}(\psi - \ln r) - (\psi - \ln r)\frac{\partial\phi}{\partial n})]dl$$
$$(3.94)$$

Notice that in the above equation both ϕ and $\frac{\partial\phi}{\partial n}$ are present in the integrand. However, in the Dirichlet problem only ϕ is prescribed on the boundary. So we choose the harmonic function ψ in such a way as to ensure that a function G defined by the difference

$$G(\xi, \eta; x, y) = \psi(\xi, \eta; x, y) - \ln r \qquad (3.95)$$

vanishes on $\partial\Omega$:

$$G(\vec{\xi}, \vec{x}) = 0 \quad \text{on} \quad \partial\Omega \qquad (3.96)$$

where $\vec{\xi} = (\xi, \eta)$ and $\vec{x} = (x, y)$. G is called the Dirichlet Green's function associated with the two dimensional Laplacian operator for the region Ω. G satisfies the Laplace's equation for $\vec{\xi} \neq \vec{x}$. (3.96) gives the boundary condition $G = 0$ on $\partial\Omega$.

In terms of G, (3.94) reads

$$\phi(P) = -\frac{1}{2\pi} \int \int_{\Omega} G(Q; P) \nabla^2 \phi(Q) d\Omega - \frac{1}{2\pi} \int_{\partial\Omega} \phi(Q) \frac{\partial}{\partial n} G(Q; P) dl \qquad (3.97)$$

Like ψ, if ϕ is also a two-dimensional harmonic function a further reduction for $\phi(P)$ is possible

$$\phi(P) = -\frac{1}{2\pi} \int_{\partial\Omega} \phi(Q) \frac{\partial}{\partial n} G(Q; P) dl \qquad (3.98)$$

Thus it is reasonable to claim from the above relation that $\phi(P)$ represented in the form

$$\phi(P) = -\frac{1}{2\pi} \int_{\partial\Omega} f(x, y) \frac{\partial}{\partial n} G(Q; P) dl \qquad (3.99)$$

is a solution of the two-dimensional Dirichlet problem for the Lapalce's equation namely,

$$\left(\frac{\partial^2}{\partial x^2} + \frac{\partial^2}{\partial y^2}\right) \phi(x, y) = 0 \quad \text{on} \quad \Omega \qquad (3.100)$$

where

$$\phi = f(x, y) \quad \text{on} \quad \partial\Omega \qquad (3.101)$$

To establish that (3.99) is a plausible solution of the Dirichlet problem, it is necessary to show that the Green's function G exists and further that $\phi(\vec{x})$ approaches the limit $f(\vec{x})$ as any point $\vec{x}$ in the two-dimensional region Ω tends to its surface value. The demonstration of this proposition is omitted here.

In the three dimensional case we have in place of (3.97) the result

$$\phi(x,y,z) = -\frac{1}{4\pi} \int\int\int_\Omega (\psi + \frac{1}{r})\nabla^2\phi d\Omega - \frac{1}{4\pi}\int\int_{\partial\Omega}[(\phi\frac{\partial}{\partial n}(\psi + \frac{1}{r}) - (\psi + \frac{1}{r})\frac{\partial\phi}{\partial n})]dl$$
$$(3.102)$$

where ψ is a harmonic function Ω, $r^2 = (x-\xi)^2 + (y-\eta)^2 + (z-\zeta)^2$ and (ξ,η,ζ) refers to the variable point $Q \in \Omega$. Here $\frac{1}{r}$ replaces $\ln r$ because $\frac{1}{r}$ is a fundamental solution of the Laplace's equation in three-dimensions.

Choosing ϕ in such a way that the Green's function G defined by

$$G(\xi,\eta,\zeta;x,y,z) = \psi(\xi,\eta,\zeta;x,y,z) + \frac{1}{r} \qquad (3.103)$$

vanishes on $\partial\Omega$ similar to (3.96) , we obtain for $\phi(P)$

$$\phi(P) = -\frac{1}{4\pi}\int\int_\Omega G(Q;P)\nabla^2\phi(Q)d\Omega - \frac{1}{4\pi}\int_{\partial\Omega}\phi(Q)\frac{\partial}{\partial n}G(Q;P)dl \qquad (3.104)$$

Like ψ, if ϕ is also a three-dimensional harmonic function we get the reduced form for $\phi(P)$

$$\phi(P) = -\frac{1}{4\pi}\int_{\partial\Omega}\phi(Q)\frac{\partial}{\partial n}G(Q;P)dl \qquad (3.105)$$

We can therefore reasonably claim that $\phi(P)$ as given by

$$\phi(P) = -\frac{1}{4\pi}\int_{\partial\Omega}f(\vec{\xi})\frac{\partial}{\partial n}G(\vec{\xi};\vec{x})dl \qquad (3.106)$$

is a solution of the three-dimensional Dirichlet problem namely,

$$(\frac{\partial^2}{\partial x^2} + \frac{\partial^2}{\partial y^2} + \frac{\partial^2}{\partial z^2})\phi(x,y,z) = 0 \quad \text{on} \quad \Omega \qquad (3.107)$$

where

$$\phi = f(x,y,z) \quad \text{on} \quad \partial\Omega \qquad (3.108)$$

As with the two-dimensional case, to establish that (3.106) is a plausible solution of the Dirichlet problem, it is necessary to show that the Green's function G exists and further that $\phi(\vec{x})$ approaches the limit $f(\vec{x})$ as any point $\vec{x}$ in the three-dimensional region Ω approaches the surface $\partial\Omega$. The demonstration of this proposition is omitted here.

We now move to some applications of the above results.

3.7   Dirichlet problem for circle, sphere and half-space

(a) Circle

Consider a circle C of radius a, centred at O as shown in Figure 3.1. The points Q and P are taken, as usual, to be the variable point and fixed point respectively. In polar form we take their coordinates to be $Q = (\rho, \theta)$ and $P = (r, \theta')$. We adopt a geometrical approach and to this end we identify P' as the inverse point of P with respect to the circle C. We denote the distances as

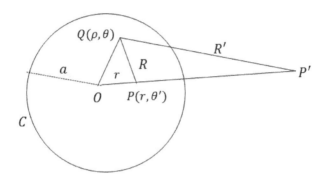

FIGURE 3.1: Circle C of radius a.

$$PQ = R, \quad P'Q - R', \quad OP = r \tag{3.109}$$

We address the Dirichlet problem for the two-dimensional Laplace's equation

$$(\frac{\partial^2}{\partial x^2} + \frac{\partial^2}{\partial y^2})\phi(x, y) = 0 \tag{3.110}$$

in the region Ω that is bounded by the circle C. The boundary condition is provided by

$$\phi = f(\theta) \quad \text{on} \quad r = a \tag{3.111}$$

where $r = \sqrt{x^2 + y^2}$ and $\theta \in (0, 2\pi]$.

We take the following choice of ψ to make use of the Green's function as noted in (3.95)

$$\psi = \ln(\frac{rR'}{a}) \tag{3.112}$$

It works from the geometrical point of view. The reason is that if Q is any point on the boundary of the circle, the triangles POQ and $P'OQ$ become similar and so P' as an inverse point of P leads to

$$OP \cdot OP' = a^2 \quad \Rightarrow \quad \frac{OP}{OQ} = \frac{OQ}{OP'} \tag{3.113}$$

The above equality signals that the angles OQP and $OP'Q$ are equal. Therefore we can write down

$$\frac{OQ}{OP} = \frac{R'}{R} = \frac{a}{r} \tag{3.114}$$

Thus the ratio $\frac{rR'}{aR} = 1$ and hence its log vanishes. An implication is that if the point Q is on the boundary then the Green's function given by

$$G(Q, P) = \ln(\frac{1}{R}) - \ln(\frac{a}{rR'}) = \ln(\frac{rR'}{aR}) \tag{3.115}$$

vanishes. (3.115) is the desired form of the Green's function for the problem of the circle

From Figure 3.1 it is clear that the angle $\angle QOP = \theta - \theta'$. Further, the following trigonometrical equalities hold

$$R^2 = r^2 + \rho^2 - 2r\rho\cos(\theta - \theta'), \quad R'^2 = \frac{1}{r^2}[a^4 + r^2\rho^2 - 2a^2r\rho\cos(\theta - \theta')] \tag{3.116}$$

So substituting these expressions in (3.115) we obtain the corresponding expression for the Green's function

$$G(Q, P) = \frac{1}{2}\ln\frac{[a^4 + r^2\rho^2 - 2a^2r\rho\cos(\theta - \theta')]}{[a^2(r^2 + \rho^2 - 2r\rho\cos(\theta - \theta'))]} \tag{3.117}$$

On C, the derivative $\frac{\partial G(Q,P)}{\partial n}$ assumes the form

$$\frac{\partial G(Q, P)}{\partial n}\Big|_C = \frac{\partial G(Q, P)}{\partial n}\Big|_{\rho=a} = \frac{r^2 - a^2}{a[r^2 + a^2 - 2ra\cos(\theta - \theta')]} \tag{3.118}$$

We already have shown that if ϕ is harmonic in a region Ω bounded by a closed curve then at any point in Ω, ϕ is given by (3.98). In the present case, $\phi = f(\theta)$ on the circle C and so on using (3.118) it turns out to be

$$\phi(r) = \frac{a^2 - r^2}{2\pi}\int_0^{2\pi}\frac{f(a, \theta)d\theta}{r^2 + a^2 - 2ra\cos(\theta - \theta')} \tag{3.119}$$

The above formula is called Poisson integral for the disc or the complete circle and solves the Dirichlet's problem, summarized in (3.110) and (3.111), in the interior of the circle.

It is simple to deduce from the above formula the value of ϕ at the point $r = 0$

$$\phi(0) = \frac{1}{2\pi} \int_0^{2\pi} f(\theta)d\theta \tag{3.120}$$

We see that the value of ϕ at the centre of the circle is equal to the average value of f on the boundary.

We leave as an exercise the proof of the following equality:

$$\frac{a^2 - r^2}{2\pi} \int_0^{2\pi} \frac{d\theta}{r^2 + a^2 - 2ra\cos(\theta - \theta')} = 1 \tag{3.121}$$

(b) Sphere

We now proceed to determine the Green's function for the Dirichlet problem in the interior of the sphere. Consider a sphere S of radius a, centred at O. The points Q and P correspond to the variable point and fixed point respectively. We assign their spherical polar coordinates to be $Q = (\rho, \theta, \lambda)$ and $P = (r, \theta', \lambda')$ respectively. As in the case of the circle we take recourse to a geometrical analysis of the problem. Let P' be the inverse point of P with respect to the sphere S as shown in Figure 3.2. We denote the distances as

$$PQ = R, \quad P'Q - R', \quad OP = r \tag{3.122}$$

We consider the Dirichlet problem for the three-dimensional Laplace's equation

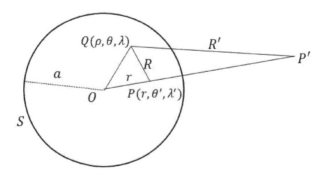

FIGURE 3.2: Sphere S of radius a.

$$\left(\frac{\partial^2}{\partial x^2} + \frac{\partial^2}{\partial y^2} + \frac{\partial^2}{\partial z^2}\right)\phi(x, y, z) = 0 \tag{3.123}$$

in the region Ω bounded by the sphere S. The boundary condition for ϕ is given by

$$\phi = f(\theta, \lambda) \quad \text{on} \quad r = a \tag{3.124}$$

where $r = \sqrt{x^2 + y^2 + z^2}$, $\theta \in (0, \pi]$ and $\lambda \in (0, 2\pi]$.

To facilitate the use of Green's function which for the three-dimensional case is given by (3.103) we consider the following choice

$$\psi = -\frac{a}{rR'} \tag{3.125}$$

This ensures that ψ is harmonic.

As in the circle's case we see that if Q is any point on the boundary of the sphere, the triangles POQ and $P'OQ$ become similar and so P' as an inverse point of P leads to

$$OP \cdot OP' = a^2 \quad \Rightarrow \quad \frac{OP}{OQ} = \frac{OQ}{OP'} \tag{3.126}$$

indicating that the angles OQP and $OP'Q$ are equal. Therefore we can write down

$$\frac{OQ}{OP} = \frac{QP'}{QP} = \frac{R'}{R} = \frac{a}{r} \tag{3.127}$$

Thus the ratio $\frac{rR'}{aR} = 1$ implying that if the point Q is on the boundary then the Green's function given by the form

$$G(Q, P) = \frac{1}{R} - \frac{a}{rR'} \tag{3.128}$$

vanishes. (3.128) gives the general form of the Green's function for the problem of sphere.

From Figure 3.2 it is clear that the following equalities hold

$$R^2 = r^2 + \rho^2 - 2r\rho \cos\alpha, \quad R'^2 = \frac{1}{r^2}[a^4 + r^2\rho^2 - 2a^2 r\rho \cos\alpha] \tag{3.129}$$

where $\cos\alpha$ is

$$\cos\alpha = \cos\theta \cos\theta' + \sin\theta \sin\theta' \cos(\lambda - \lambda') \tag{3.130}$$

So substituting these expressions in the expression of the Green's function (3.128) we obtain

$$G(Q, P) = (r^2 + \rho^2 - 2r\rho \cos\alpha)^{-\frac{1}{2}} - a(a^4 + r^2\rho^2 - 2a^2 r\rho \cos\alpha)^{-\frac{1}{2}} \tag{3.131}$$

On S, the derivative $\frac{\partial G(Q,P)}{\partial n}$ is given by

$$\frac{\partial G(Q,P)}{\partial n}\Big|_S = \frac{\partial G(Q,P)}{\partial n}\Big|_{\rho=a} = \frac{r^2 - a^2}{a[r^2 + a^2 - 2ra\cos\alpha]^{\frac{3}{2}}} \tag{3.132}$$

We have already shown that if ϕ is harmonic in a region Ω bounded by a closed surface then at any point in Ω, ϕ is given by (3.106). In the present case $\phi(P)$ turns out to be

$$\phi(P) = \frac{a(a^2 - r^2)}{4\pi} \int_0^{2\pi} \int_0^\pi \frac{f(a,\theta,\lambda)\sin\theta d\theta d\lambda}{(r^2 + a^2 - 2ra\cos\alpha)^{\frac{3}{2}}} \tag{3.133}$$

on using (3.132). The above formula is the Poisson integral for the sphere and solves the Dirichlet's problem summarized in (3.123) and (3.124).

(c) Half-space

Here the Dirichlet's problem seeks to find the solution of the Laplace's equation

$$\left(\frac{\partial^2}{\partial x^2} + \frac{\partial^2}{\partial y^2} + \frac{\partial^2}{\partial z^2}\right)\phi(x,y,z) = 0 \tag{3.134}$$

in the half-space region $z \geq 0$. The boundary condition is given on the plane $z = 0$ namely

$$\phi = f(x,y), \quad \text{on} \quad z = 0 \tag{3.135}$$

The function f is assumed to be twice differentiable with each derivative being continuous.

In Figure 3.3, the points $Q(\xi,\eta,\zeta)$ and $P(x,y,z)$ correspond to the variable point and fixed point respectively in the region $z \geq 0$. Let $P'(x,y,-z)$ be the image of P with respect to the plane $z - 0$. We denote the distances as

$$PQ = R, \quad P'Q = R' \tag{3.136}$$

In three-dimensions the Green's function reads from (3.103)

$$G(Q;P) = \frac{1}{R} + \psi \tag{3.137}$$

where in the present case $r = R$ and ψ is harmonic in $z \geq 0$ i.e.

$$\left(\frac{\partial^2}{\partial \xi^2} + \frac{\partial^2}{\partial \eta^2} + \frac{\partial^2}{\partial \eta^2}\right)\psi(x,y,z) = 0, \quad z \geq 0 \tag{3.138}$$

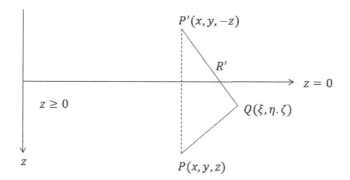

FIGURE 3.3: The half space in the region $z \geq 0$.

Taking $\psi = -\frac{1}{R'}$ the relevant Green's function is

$$G(Q,P) = \frac{1}{R} - \frac{1}{R'} \tag{3.139}$$

This is so because if the point Q is on the boundary $z = 0$ then from Figure 3.3, it is clear that $R = R'$ implying $G(Q,P) = \frac{1}{R} - \frac{1}{R'} = 0$.

The expressions for R and R' are

$$R^2 = (\xi - x)^2 + (\eta - y)^2 + (\zeta - z)^2, \quad R'^2 = (\xi - x)^2 + (\eta - y)^2 + (\zeta + z)^2 \tag{3.140}$$

and the normal derivative of $G(Q; P)$ is given by

$$\frac{\partial G(Q,P)}{\partial n}\Big|_{\zeta=0} = -\frac{\partial G(Q,P)}{\partial \zeta}\Big|_{\zeta=0} = \frac{1}{R^2}\frac{\partial R}{\partial \zeta}\Big|_{\zeta=0} - \frac{1}{R'^2}\frac{\partial R'}{\partial \zeta}\Big|_{\zeta=0} \tag{3.141}$$

A little algebra yields

$$\frac{\partial G(Q,P)}{\partial n}\Big|_{\zeta=0} = -\frac{2z}{[(\xi-x)^2 + (\eta-y)^2 + z^2]^{\frac{3}{2}}} \tag{3.142}$$

We already know that if ϕ is harmonic in a region Ω bounded by a closed surface then at any point in Ω, ϕ is given by (3.106). In the present case this gives for $\phi(P)$

$$\phi(P) = \frac{z}{2\pi} \int_{-\infty}^{\infty} \int_{-\infty}^{\infty} \frac{f(\xi,\eta)d\xi d\eta}{[(\xi-x)^2 + (\eta-y)^2 + z^2]^{\frac{3}{2}}} \tag{3.143}$$

which solves the Dirichlet's problem posed by (3.134) and (3.135).

3.8   Summary

To summarize, we introduced in this chapter the elliptic form of PDE. We described the method of solutions through the separation of variables touching upon the plane polar coordinates, spherical polar coordinates and cylindrical coordinates. We discussed the harmonic function and gave a formal proof of Gauss' mean value theorem. We also furnished the proofs of the maximum-minimum principle for Poisson's and Laplace's equations and commented on the existence and uniqueness of the solutions for these equations. Subsequently we addressed the normally directed distribution of doublets and outlined the procedure of generating Green's function for the Laplacian operator both in two and three dimensions. As applications we derived the solutions of the Dirichlet problem for the circle, the sphere and the half-space.

Exercises

1. Solve Laplace's equation

$$\phi_{xx}(x, y) + \phi_{yy}(x, y) = 0$$

subject to the boundary conditions

$$\phi(x, 0) = \sin(\pi x), \quad \phi(x, 1) = \sin(\pi x)e^{-\pi}, \quad \phi(0, y) = 0, \quad \phi(1, y) = 0$$

and show that the solution is $\phi(x, y) = \sin(\pi x)e^{-\pi y}$.

2. Show that $\phi(x, y) = \frac{1}{6}x^4 - x^2 y^2 + \frac{1}{6}y^4$ is a harmonic function.

3. The planes $x = 0, x = \pi$ and $y = 0$ form the boundary of a semi-infinite two-dimensional space. The potential ϕ satisfies the Poisson's equation

$$\phi_{xx}(x, y) + \phi_{yy}(x, y) = -\rho$$

in the presence of a uniform distribution of charge density $\frac{\rho}{4\pi}$. While ϕ is bounded over the concerned region, the boundary conditions are

$$\phi(0, y) = 0, \quad \phi(\pi, y) = 1, \quad \phi(x, 0) = 0$$

Show that $\phi(x, y)$ is given by the form

$$\phi(x, y) = \frac{2}{\pi} \sum_{n=1}^{\infty} \left[\frac{\rho((-1)^n - 1)}{n^3} + \frac{(-)^n}{n} \right] (e^{-ny} - 1) \sin(nx)$$

4. Consider the Beltrami equations, which are a generalization of Cauchy-Riemann equations, as given by the pair of functions ϕ and ψ satisfying the equations

$$\zeta \phi_x - b\psi_x - c\psi_y = 0 \quad \zeta \phi_y + a\psi_x + b\psi_y = 0$$

where $\zeta \neq 0$, the matrix

$$M = \begin{bmatrix} a & b \\ c & d \end{bmatrix}$$

is positive definite and the coefficients a, b, c and d are given functions of x and y. Show that such a system is elliptic.

5. Show that $\frac{\sin(k|x|)}{|x|}$ solves the Helmholtz's equation $\triangle \phi + k^2 \phi = 0$ in $\Re^n$ in the deleted neighbourhood of the origin.

6. By analyzing the coefficient matrix of the following PDE

$$\phi_{xx} - \phi_{xy} + \phi_{yy} = f(x, y)$$

show that the equation is of elliptic type.

7. Establish that a bounded harmonic function in $\Re^n$ is a constant.

8. Consider the Lapalce's equation $\phi_{xx}(x, y) + \phi_{yy}(x, y) = 0$ in the region $x^2 + y^2 < 1$ along with the boundary condition $\phi = xye^{x^2+y^4+2}$ for $x^2 + y^2 = 1$. Show by the mean value theorem that $\phi(0, 0) = 2$.

9. Solve Laplace's equation

$$\phi_{xx}(x, y) + \phi_{yy}(x, y) = 0, \quad 0 < y < \infty$$

with $\phi = f(x)$ on $y = 0$ and show that the solution can be expressed as

$$\phi(x, y) = \frac{y}{\pi} \int_0^\infty \frac{f(\xi)d\xi}{(x - \xi)^2 + y^2}$$

10. In the semi-annulus region defined by $r \in (1, 2)$ and $\theta \in (0, \pi)$ show that the two-dimensional Lapalce's equation

$$\frac{1}{r} \frac{\partial}{\partial r} (r \frac{\partial \phi}{\partial r}) + \frac{1}{r^2} \frac{\partial^2 \phi}{\partial \theta^2} = 0$$

subject to the boundary conditions

$$\phi(r, 0) = 0 = \phi(r, \pi) \quad \text{in} \quad 1 < r < 2$$

and

$$\phi(1, \theta) = \sin \theta, \quad \phi(2, \theta) = 0 \quad \text{in} \quad 0 < \theta < \pi$$

has the solution $\phi(r, \theta) = (\frac{4}{3r} - \frac{r}{3}) \sin \theta$.

Chapter 4

PDE: Hyperbolic form

The wave equation which was introduced in Chapter 1 is a prototypical example for the class of hyperbolic PDEs. It arises in many branches of physics. A quick review of the derivation of the wave equation goes as follows. Consider a region of surface S enclosing a volume V. Newton's force-acceleration relation gives on integration

$$\frac{\partial^2}{\partial t^2} \int_V \phi(\vec{r}, t) dv = -\int_S \vec{F}(\vec{r}, t) \cdot \vec{n} ds$$

where $\vec{n}$ is the outward drawn unit normal to S and we have set the mass-density factor to be unity. Employing Gauss's divergence theorem

$$\int_V div \vec{F}(\vec{r}, t) dv = \int_S \vec{F}(\vec{r}, t) \cdot \vec{n} ds$$

allows us to re-cast the previous equation to the form

$$\int_V div \vec{F}(\vec{r}, t) dv = -\frac{\partial^2}{\partial t^2} \int_V \phi(\vec{r}, t) dv = -\int_V \frac{\partial^2 \phi(\vec{r}, t)}{\partial t^2} dv$$

which implies

$$div \vec{F}(\vec{r}, t) = -\frac{\partial^2 \phi(\vec{r}, t)}{\partial t^2}$$

Taking F to be proportional to the gradient of ϕ but acting in the opposite direction i.e. given by $\vec{F}(\vec{r}, t) = -c^2 \nabla \phi(\vec{r})$, where c is a constant we then have the wave equation

$$\phi_{tt} - c^2 \nabla^2 \phi = 0$$

where c is the propagation speed.

The above equation is obtained in the homogeneous form. A non-homogeneous equation of course is more general where one can have a term on the right side as a function of the spatial variables as in the case of Poisson equation. If we consider the problem of small oscillations about some stationary or equilibrium point then in one-dimension the wave equation stands for the displacement from the stationary position. In two-dimensions the wave equation is relevant to the vibrations of a membrane. In three dimensions the wave equation appears in electromagnetic theory where ϕ could represent a component of say the electric field $\tilde{\mathbf{E}}$. Wave equations also play important roles in theory of hydromechanics, optical problems, heat transfer etc.

In this chapter we are going to concentrate on the wave equation as the typical representative equation of the PDE in the hyperbolic form and study some of its different properties. We begin, first of all, by taking up the derivation of D'Alembert's solution by focusing on the one-dimensional wave equation.

4.1   D'Alembert's solution

The one-dimensional wave equation is categorized as a classical example for studying travelling-wave phenomena. It is also relevant in accounting for the transverse displacements of a vibrating string.

We will be interested in solving the Cauchy problem defined on an infinite domain and towards this pursuit we will consider the following one-dimensional form

$$\phi_{tt} - c^2 \phi_{xx} = 0, \quad -\infty < x < +\infty, \quad t \geq 0 \tag{4.1}$$

being subjected to a pair of initial conditions on the position and velocity

$$\phi(x, 0) = f(x), \quad \phi_t(x, 0) = g(x) \tag{4.2}$$

We assume that the function $f(x)$ is twice differentiable and that $g(x)$ is differentiable.

We will determine the solution in some domain of (t, x) plane. Setting

$$\xi = x - ct, \quad \eta = x + ct \tag{4.3}$$

the PDE (4.1) is readily converted to the mixed form

$$\frac{\partial^2 \phi}{\partial \xi \, \partial \eta} = 0 \tag{4.4}$$

Equation (4.4) has the immediate solution

$$\phi = \chi(\xi) + \psi(\eta) = \chi(x - ct) + \psi(x + ct) \tag{4.5}$$

Let us recall from Chapter 1 that the families of straight lines $\xi = x - ct =$ constant and $\eta = x + ct =$ constant are referred to as the characteristics. It is across them that a solution can have jump discontinuities. If ϕ is given on the segments of two of the characteristics and if we complete a rectangle by drawing the parallel lines to $\xi =$ constant and $\eta =$ constant, then any discontinuity on the characteristics will be propagated into the interior of the rectangle. The functions χ and ψ represent respectively a right- and left-travelling wave with velocity c.

Employing the given Cauchy conditions we obtain for ϕ and ψ the expressions

$$\chi(x) = \frac{1}{2}[f(x) - \int_{x_0}^{x} g(s) \, ds - k] \tag{4.6}$$

and

$$\psi(x) = \frac{1}{2}[f(x) + \int_{x_0}^{x} g(s) \, ds + k] \tag{4.7}$$

where x_0 is arbitrary and k is a constant of integration. Inserting (4.6) and (4.7) in (4.5) we obtain

$$\phi(x,t) = \frac{1}{2}[f(x - ct) + f(x + ct) + \int_{x-ct}^{x+ct} g(s) \, ds] \tag{4.8}$$

which is called the D'Alembert's solution.

Some remarks are in order about the energy equation for the infinite string for which we got the D'Alembert's solution. Identifying for the kinetic energy the term ϕ_t^2 and for the potential energy the term $c^2 \phi_x^2$ we can readily write down for the energy E the integral

$$E = \frac{1}{2} \int_{-\infty}^{+\infty} (\phi_t^2 + c^2 \phi_x^2) dx \tag{4.9}$$

where we assume ϕ_t and ϕ_x to decay rapidly and asymptotically. Note that we set the mass density to be equal to unity. Taking the time derivative

$$\frac{dE}{dt} = \int_{-\infty}^{+\infty} (\phi_t \phi_{tt} + c^2 \phi_x \phi_{xt}) dx \tag{4.10}$$

gives on integrating by parts

$$\frac{dE}{dt} = \int_{-\infty}^{+\infty} \phi_t(\phi_{tt} - c^2 \phi_{xx}) dx = 0 \tag{4.11}$$

where we have dropped the boundary term because of the decaying condition at infinity and used the wave equation. (4.11) reveals that the energy is conserved.

From a physical point of view it is clear that the given set of Cauchy initial conditions (4.2) are appropriate for the uniqueness of the solution (4.8). Indeed if we take two solutions labeled by ϕ_1 and ϕ_2 with the corresponding $f = f_1$ and f_2 and $g = g_1$ and g_2 for the initial conditions, then the difference $\Phi(x,t) \equiv \phi_1 - \phi_2$ satisfies the one-dimensional wave equation with $\Phi(x,0) = 0$ and $\Phi_t(x,0) = 0$. It shows that the energy for Φ at $t = 0$ is zero and stays so for all time by the conservation of the energy proved above. Further $\Phi(x,0) = 0$ and hence $\Phi \equiv 0$ justifying the uniqueness of the solution obtained in (4.8).

That the D'Alembert's solution is well-posed can be established by considering two solutions such that their difference is small $: |f_1 - f_2| < \epsilon$ and $|g_1 - g_2| < \epsilon$. D'Alembert's formula (4.8) gives

$$|\phi_1 - \phi_2| < \frac{1}{2}(\epsilon + \epsilon) + \frac{1}{2c}\int_{x-ct}^{x+ct} \epsilon \, ds = \epsilon(1 + t) \tag{4.12}$$

We therefore see that for a finite time span, a small change in the initial conditions produces a proportionately small change in the solution signifying that the solution is well posed.

Consider an arbitrary point $P(\overline{x}, c\overline{t})$ in the $xt-$plane as shown in Figure 4.1. The characteristics through P are given by the equation $x - \overline{x} = \pm c(t - \overline{t})$ which intersects the x axis at A and B whose abscissae are respectively $\overline{x} - c\overline{t}$ and $\overline{x} + c\overline{t}$. Thus we see that D'Alembert's solution at P depends on f at A and B and g between A and B. The closed segment AB, which is also the base of an isosceles triangle or the characteristic triangle PAB, is called the domain of dependence of P.

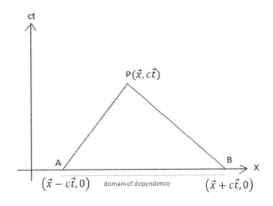

FIGURE 4.1: Domain of dependence.

Example 4.1

Construct the solution of the following wave equation

$$\phi_{tt} = c^2 \phi_{xx}, \quad x > 0, \quad t > 0$$

subject to the following initial-boundary conditions

$$\phi(x,0) = 0 = \phi_t(x,0), \quad x > 0, \quad \phi(0,t) = s(t), \quad t > 0$$

Demonstrate that the solution can be expressed in the form

$$\phi(x,t) = \begin{cases} 0 & \text{if} \quad 0 < t \le \frac{x}{c} \\ s(t - \frac{x}{c}) & \text{if} \quad t > \frac{x}{c} \end{cases} \tag{4.13}$$

4.2 Solving by Riemann method

We take the following normal form of a hyperbolic PDE

$$L(\phi) \equiv \phi_{xy} + D\phi_x + E\phi_y + F\phi = 0 \tag{4.14}$$

where D, E and F are functions of the variables x and y. The adjoint to L from (2.78) is

$$M(\psi) \equiv \psi_{xy} - (D\psi)_x - (E\psi)_y + F\psi = 0 \tag{4.15}$$

The essence of Riemann method consists in exploiting the two-dimensional form of Gauss' theorem according to which one equates the flux integral of a vector $\vec{A}$ through the positively oriented closed boundary curve C with the double integral of the divergence of $\vec{A}$ over the full region S namely, $\int_C A_{\hat{n}} dl = \int \int_S div \vec{A} ds$, where $\hat{n}$ is an outwardly drawn normal to C. Actually this divergence form of Gauss' theorem is the counterpart of Green's theorem for the curl. Indeed if the components of $\vec{A}$ are (P, Q) then the divergence theorem assumes the form $\oint_C Pdy - Qdx = \int \int_S (\frac{\partial P}{\partial x} + \frac{\partial Q}{\partial y})$.

Since we can write

$$\psi L\phi - \phi M\psi = \frac{\partial P}{\partial x} + \frac{\partial Q}{\partial y} \tag{4.16}$$

employing the two-dimensional divergence theorem we connect

$$\int_S [\psi L\phi - \phi M\psi] ds = \int_S (\frac{\partial P}{\partial x} + \frac{\partial Q}{\partial y}) ds = \int_C [P\cos(n,x) + Q\cos(n,y)] dl \tag{4.17}$$

where from the given expressions of L and M

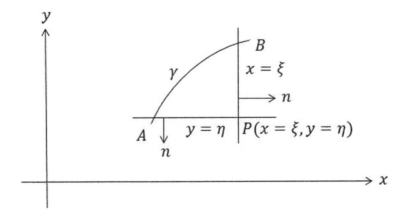

FIGURE 4.2: The curve γ in the region S.

$$P = \frac{1}{2}(\psi\phi_y - \phi\psi_y) + D\phi\psi, \quad Q = \frac{1}{2}(\psi\phi_x - \phi\psi_x) + E\phi\psi \qquad (4.18)$$

We determine ψ from the solution of the homogeneous equation

$$M(\psi) = 0 \qquad (4.19)$$

subject to the requirement that

$$\psi = 1 \quad \text{at} \quad P(x = \xi, y = \eta) \qquad (4.20)$$

and that on the characteristics it obeys the following conditions

$$\psi_y - D\psi = 0 \quad \text{on} \quad x = \xi \quad \text{and} \quad \psi_x - E\psi = 0 \quad \text{on} \quad y = \eta \qquad (4.21)$$

We refer to ψ as the Riemann function or characteristic function. Riemann's method is essentially a reduction of the problem to a more easily tractable boundary-value problem proposed in (4.18), (4.19) and (4.20).

To proceed with the relation (4.16), we look at the Figure 4.2 in which a region is formed when the non-charateristic curve γ is intersected by the hands of the characteristics AP and PB at the points A and B. We take the coordinates of the point P to be (ξ, η), with $y = \eta$ and $x = \xi$ being the equations of the lines AP and PB respectively. We assume ϕ and $\frac{\partial \phi}{\partial n}$ to be given on γ. Employing (4.18) and noting that $L(\phi) = 0$ is the given PDE, we obtain on integrating over the periphery of the region bounded by the curve γ and the lines PB and PA

$$\int_{(\gamma + PB + PA)} [P\cos(n,x) + Q\cos(n,y)]dl = 0$$

where P and Q are given by (4.17).

In the integration over PB, only the P-term survives because $\cos(n,y) = 0$. Writing

$$\frac{1}{2}\int_P^B \psi\phi_y\,dy = \frac{1}{2}\phi\psi|_P^B - \frac{1}{2}\int_P^B \phi\psi_y\,dy$$

we obtain

$$\int_P^B P\,dy = \frac{1}{2}[(\phi\psi)|_B - (\phi\psi)|_P] - \int_P^B \phi(\psi_y - D\psi)dy \qquad (4.22)$$

For the integral $\int_A^P$, we notice that $\cos(n,x) = 0$ and $\cos(n,y) = -1$, n being the outer normal, we obtain

$$-\int_A^P Q\,dx = \frac{1}{2}[(\phi\psi)|_A - (\phi\psi)|_P] + \int_A^P \phi(\psi_x - E\psi)dx \qquad (4.23)$$

The two integrals in the right sides of (4.21) and (4.22) vanish because of the conditions (4.20) that the conditions fulfill. Further, making use of (4.19) we arrive at the central result of the Riemann's method

$$\phi_P = \int_\gamma [P\cos(n,x) + Q\cos(n,y)]dl + \frac{1}{2}[(\phi\psi)|_A + (\phi\psi)_B] \qquad (4.24)$$

giving the value of ϕ at an arbitrary point P.

To see how Riemann's method works in practice we consider an example from hydro-mechanics where it is given that

$$D = -\frac{\lambda}{x+y} = E, \quad F = 0$$

where λ is a constant.

Substituting in the two PDEs written down in (4.20) we immediately obtain two solutions for ψ

$$x = \xi: \quad \psi = c_1(\xi + y)^{-\lambda}; \quad y = \eta: \quad \psi = c_2(x + \eta)^{-\lambda} \qquad (4.25)$$

where c_1 and c_2 are constants.

Requiring the condition (4.19) to hold implies $c_1 = c_2 = (\xi + \eta)^\lambda$ which means

$$\psi = (\frac{\xi + \eta}{x + y})^\lambda \qquad (4.26)$$

However such a ψ does not provide a general solution to $M(\psi) = 0$ which is given by a more complicated form

$$\psi = (\frac{\xi + \eta}{x + y})^\lambda \; {}_2F_1(\lambda + 1, -\lambda, 1, ; z) \qquad (4.27)$$

where $_2F_1(\lambda + 1, -\lambda, 1, ; z)$ is the Gauss hypergeometric series and the quantity z is defined to be

$$z = -\frac{(x - \xi)(y - \eta)}{(x + y)(\xi + \eta)} \tag{4.28}$$

Proof: We briefly outline the steps. Taking the form (4.27) for ψ we compute first the derivatives ψ_x, ψ_y and ψ_{xy}. Substituting them in the equation for $M(\psi) = 0$ which has the form

$$M(\psi) \equiv \psi_{xy} + \frac{\lambda}{x+y}(\psi_x + \psi_y) - 2\lambda\frac{\psi}{(x+y)^2} = 0 \tag{4.29}$$

results in

$$z_x z_y F''(z) + z_{xy} F'(z) - \frac{\lambda(\lambda+1)}{(x+y)2} F = 0 \tag{4.30}$$

where $_2F_1(\lambda + 1, -\lambda, 1, ; z) \equiv F(z)$. Using (4.28) we can now easily convert (4.30) into the standard differential equation $z(1 - z)F''(z) + (1 - 2z)F' + (\lambda(\lambda + 1)F = 0$.

Example 4.2

Find the Riemann function when L is given in the form

$$L\phi \equiv \phi_{xy} + \lambda\phi = \rho(x, y)$$

where λ is a constant and ρ is the inhomogeneous term.

The Riemann function satisfies the adjoint equation

$$M\psi \equiv \psi_{xy} + \lambda\psi = 0$$

along with $\psi_x = 0$ on $y = \eta$, $\psi_y = 0$ on $x = \xi$ and $\psi(x = \xi, y = \eta) = 1$.

A plausible form for ψ is

$$\psi = \chi(t), \quad t = (x - \xi)(y - \eta)$$

Substituting in the above equation for ψ gives the differential equation

$$t\frac{d^2\chi}{dt^2} + \frac{d\chi}{dt} + \lambda\chi = 0$$

A change of variable $z = 2\sqrt{\lambda t}$ yields the form

$$\frac{d^2\chi}{dz^2} + \frac{1}{r}\frac{d\chi}{dz} + \chi = 0$$

which can be recognized as the Bessel equation of order zero. We see that the solution

$$\chi = J_0(z), \quad \chi = J_0[2\sqrt{\lambda(x-\xi)(y-\eta)}]$$

satisfies $\chi = 1$ at $t = 0$ i.e. $z = 0$ and hence all requirements of ψ enlisted in (4.19)-(4.21) are met.

4.3 Method of separation of variables

(a) Three dimensions: spherical polar coordinates (r, θ, ϕ)

In this section we consider solving the wave equation through the method of separation of variables. The procedure is similar to what was adopted for the Laplace equation.

Consider the three-dimensional wave equation

$$\nabla^2 \phi - \frac{1}{c^2}\phi_{tt} = 0 \tag{4.31}$$

where the form of ∇^2 in three-dimensional spherical polar coordinates (r, θ, ϕ) has been noted in (3.26). Further in (3.27) the relations of the variables r, θ and ϕ are given in terms of the Cartesian coordinates including the ranges of r, θ and ϕ.

Let us assume that the solution of ϕ exists in the variable-separated product form

$$\phi(r, \theta, \phi, t) = f(r, \theta, \phi)T(t) \tag{4.32}$$

where f is a function of (r, θ, ϕ) and T is a function of t only.

Substituting (4.32) in (4.31) gives for f

$$\nabla^2 f + k^2 f = 0 \tag{4.33}$$

which is known as the Helmholtz equation and an ODE for T

$$\frac{d^2 T}{dt^2} + c^2 k^2 T = 0 \tag{4.34}$$

where k^2 is the separation constant and arises because t has been separated out from the remaining variables.

The general solution of (4.34) is

$$T(t) = A\cos(kct) + B\sin(kct) \tag{4.35}$$

where A and B are arbitrary constants.

We now further separate $f(r, \theta, \phi)$ in the product form

$$f(r, \theta, \phi) = R(r)g(\theta, \phi) \tag{4.36}$$

where R is a function of r only and $g(\theta, \phi)$ is a function of the variables θ and ϕ only. Substituting (4.36) in (4.33) one obtains the following equations

$$\frac{d}{dr}\left(r^2 \frac{dR}{dr}\right) + (k^2 r^2 - p)R = 0 \tag{4.37}$$

and

$$\frac{1}{\sin\theta} \frac{\partial}{\partial\theta}\left(\sin\theta \frac{\partial g}{\partial\theta}\right) + \frac{1}{\sin^2\theta} \frac{\partial^2 g}{\partial\phi^2} + pg = 0 \tag{4.38}$$

where p is a separation constant and arises because the variable r has been separated out from the remaining ones (θ, ϕ) as is clear from (4.37) and (4.38).

Finally, we separate $g(\theta, \phi)$ in the form

$$g(\theta, \phi) = s(\theta)h(\phi) \tag{4.39}$$

where s is a function of θ only and $h(\phi)$ is a function of ϕ only. Substituting in (4.38) leads to the following pair of equations

$$\sin\theta \frac{\partial}{\partial\theta}\left(\sin\theta \frac{ds}{d\theta}\right) + (p\sin^2\theta - \nu^2)s = 0 = 0 \tag{4.40}$$

and

$$\frac{d^2 h}{d\phi^2} + \nu^2 h = 0 \tag{4.41}$$

To summarize, all the variables have been separated out and we have one individual equation for each of them namely, (4.34), (4.37), (4.40) and (4.41).

Making a change of variable $U = \sqrt{r}R$ converts (4.37) to

$$r^2 \frac{d^2 U}{dr^2} + r \frac{dU}{dr} + (k^2 r^2 - p - \frac{1}{4})U = 0 \tag{4.42}$$

Substitution of $y = kr$ then gives

$$y^2 \frac{d^2 U}{dr^2} + y \frac{dU}{dy} + (y^2 - \alpha^2)U = 0, \quad \alpha^2 = p + \frac{1}{4} \tag{4.43}$$

which is a Bessel differential equation of order α.

The general solution to (4.43) can be expressed as a combination of the Bessel function J_α and Neumann function Y_α. As such the radial function $R(r)$ has the form

$$R(r) = \frac{1}{\sqrt{r}}[]CJ_\alpha(kr) + DY_\alpha(kr)] \tag{4.44}$$

where C and D are constants.

Addressing now the equation (4.41) its general solution is

$$h(\phi) = E\cos(\nu\phi) + F\sin(\nu\phi) \tag{4.45}$$

where E and F are constants.

Finally for the equation (4.40) if we make the substitutions $\mu = \cos\theta$ and $p = \beta(\beta+1)$ which implies $\alpha = \beta + \frac{1}{2}$, it can be represented by

$$\frac{d}{d\mu}[(1-\mu^2)\frac{dg}{d\mu}] + [\beta(\beta+1) - \frac{\nu^2}{1-\mu^2}]g = 0 \tag{4.46}$$

The above form is readily recognized as an associated Legendre equation whose general solution is

$$g(\mu) = GP_\beta^\nu(\mu) + HQ_\beta^\nu(\mu) \tag{4.47}$$

where G, H are constants and $P_\beta^\nu(\mu)$ and $Q_\beta^\nu(\mu)$ are the associated Legendre functions which we know to be expressible in terms of hypergeometric functions.

The general solution of the three-dimensional wave equation is given by the product of the functions $T(t)$, $R(r)$, $h(\phi)$ and $g(\mu)$ whose explicit forms are shown in (4.35), (4.44), (4.45) and (4.47) respectively involving the arbitrary constants A, B, C, D, E, F, G and H. In particular for the Helmholtz part if we look for a finite and single-valued solution in the ranges $0 \le \phi < 2\pi$ and $0 \le \theta < \pi$ on the surface of a sphere having radius a, we have to restrict $D = 0$ and $H = 0$ since logarithmic singularity affects $Y_{\beta+\frac{1}{2}}(kr)$ at $r = 0$ and $Q_\beta^\nu(\mu)$ at $\mu = \pm 1$. Single-valuedness of the functions $\sin(\nu\phi)$ and $\cos(\nu\phi)$ implies ν be a positive integer or zero. Convergence of the series for $P_\beta^\nu(\mu)$ in the range $-1 \le \mu \le +1$ restricts β to a positive integer or zero in which case the series terminates..

Putting the above arguments together and keeping in mind the transformation $Y = \sqrt{r}R$ the solution of the spherical Helmholtz equation (4.33) which is regular at the origin takes the form

$$f = r^{-\frac{1}{2}}J_{n+\frac{1}{2}}(kr)P_n^m(\mu)[E_{nm}\cos(m\phi) + F_{nm}\sin(m\phi)] \tag{4.48}$$

where $\beta = n = 0, 1, 2, ...$ and $m = 0, 1, 2, ..., n$. Further, $P_n^m(\mu) = 0$ if $m > n$. Note that the portion

$$Y_n^m(\theta, \phi) = \sum_{m=0}^{n} P_n^m(\mu)[E_{nm}\cos(m\phi) + F_{nm}\sin(m\phi)] \tag{4.49}$$

denotes the spherical harmonics of degree n, the θ-dependence arising from the definition $\mu = \cos(\theta)$.

(b) Cylindrical polar coordinates (r, θ, z)

We now inquire into the solution of the Helmholtz equation in cylindrical polar coordinates (r, θ, z), It reads

$$\frac{\partial^2 f}{\partial r^2} + \frac{1}{r}\frac{\partial f}{\partial r} + \frac{1}{r^2}\frac{\partial^2 f}{\partial \theta^2} + \frac{\partial^2 f}{\partial z^2} + k^2 f = 0 \qquad (4.50)$$

where we have used ∇^2 from (3.44). In (3.45) the relations of the variables r, θ and z are given in terms of the Cartesian coordinates including the ranges of r, θ and z.

We seek solutions of (4.50) in the variable-separated product form namely,

$$f(r, \theta, z) = h(r, \theta)g(z) \qquad (4.51)$$

where f is a function of r, θ while g is a function of z only. It is readily found that h and g obey the equations

$$\frac{\partial^2 h}{\partial r^2} + \frac{1}{r}\frac{\partial h}{\partial r} + \frac{1}{r^2}\frac{\partial^2 h}{\partial \theta^2} + p^2 h = 0 \qquad (4.52)$$

and

$$\frac{d^2 g}{dz^2} + \alpha^2 g = 0 \qquad (4.53)$$

where p^2 is a separation constant and $\alpha^2 = k^2 - p^2$.

The general solution to the equation (4.53) is provided by

$$g(z) = A\cos(\alpha z) + B\sin(\alpha z) \qquad (4.54)$$

where A and B are constants.

To tackle (4.52) we subject h to a further separation of variables

$$h(r, \theta) = R(r)\Theta(\theta) \qquad (4.55)$$

where R is a function of r only and Θ is a function of θ only.

Substituting (4.55) in (4.52) we deduce for the equation of R the form

$$\frac{d^2 R}{dr^2} + \frac{1}{r}\frac{dR}{dr} + (p^2 - \frac{\nu^2}{r^2})R = 0 \qquad (4.56)$$

which is Bessel's equation of order ν and for the equation of Θ the form

$$\frac{d^2\Theta}{d\theta^2} + \nu^2\Theta = 0 \qquad (4.57)$$

where ν^2 is a separation constant.

The general solution to (4.56) is a combination of Bessel and Neumann functions of order ν as already discussed while seeking solution of Laplace's equation in cylindrical coordinates in the previous chapter namely

$$R(r) = C J_\nu(pr) + D Y_\nu(pr) \qquad (4.58)$$

where C and D are constant, the one for (4.57) is of the periodic type

$$\Theta(\theta) = E \cos(\nu\theta) + F \sin(\nu\theta) \qquad (4.59)$$

where E and F are constants. The condition $\theta + 2\pi = \theta$ requires ν to be zero or an integer.

The general solution of the Helmholtz equation is given by the product of the functions $g(z)$, $R(r)$ and $\Theta(\theta)$ whose explicit forms are shown in (4.54), (4.58) and (4.59) respectively involving the arbitrary constants A, B, C, D, E and F.

4.4 Initial value problems

(a) Three dimensional wave equation

We analyze first the following initial value problem for the three-dimensional inhomogeneous wave equation

$$\phi_{tt} - c^2 \nabla^2 \phi(x, y, z, t) = \rho(x, y, z, t) \qquad (4.60)$$

where ρ is an inhomogeneous term. We subject (4.60) to the initial conditions

$$\phi(x, y, z, 0) = f(x, y, z), \quad \phi_t(x, y, z, 0) = g(x, y, z) \qquad (4.61)$$

where f is assumed to have continuous derivatives up to second order while only continuous derivatives of g are sufficient.

To tackle the problem we proceed in two steps. In the first step we set up a scheme in which a homogeneous counterpart of the equation (4.60) is solved subject to the inhomogeneous initial conditions (4.61). In the second step we solve the inhomogeneous equation (4.60) on which a set of homogeneous initial conditions is imposed. By the principle of superposition we then arrive at a unique solution of (4.60) being accompanied by the initial conditions (4.61).

Step 1

Let us take a sphere S of radius ct whose centre is at the point (x, y, z). Its equation is given by

$$(x - x')^2 + (y - y')^2 + (z - z')^2 = c^2 t^2 \qquad (4.62)$$

where (x', y', z') is any point on the surface. The radius of the sphere being ct, we can express the coordinates (x', y', z') as

$$x' = x + lct \equiv x + (\sin \theta \cos \phi)ct \qquad (4.63)$$

$$y' = y + mct \equiv y + (\sin \theta \sin \phi)ct \qquad (4.64)$$

$$z' = z + nct \equiv z + (\cos \theta)ct \qquad (4.65)$$

Consider the function $g(x, y, z)$. Its spherical mean or the average value over S is by definition

$$\bar{g}(x, y, z, t) = \frac{1}{4\pi c^2 t^2} \int_S g(x', y', z') ds' \qquad (4.66)$$

Since $ds' = c^2 t^2 \sin \theta d\theta \dot{d\phi}$, $\bar{g}$ reads

$$\bar{g} = \frac{1}{4\pi} \int_0^\pi \int_0^{2\pi} g(x + lct, y + mct, z + nct) \sin \theta d\theta d\phi \qquad (4.67)$$

We now prove the following assertion.

A function ψ defined by

$$\psi(x, y, z, t) = t\bar{g}(x, y, z, t) \qquad (4.68)$$

satisfies the homogeneous wave equation

$$\psi_{tt} - c^2 \nabla^2 \psi(x, y, z, t) = o \qquad (4.69)$$

obeying the initial conditions

$$\psi(x, y, z, 0) = 0, \quad \psi_t(x, y, z, 0) = g(x, y, z) \qquad (4.70)$$

Proof: Partially differentiating both sides of (4.67) with respect to t gives

$$\frac{\partial \bar{g}}{\partial t} = \frac{c}{4\pi} \int_0^\pi \int_0^{2\pi} [l\frac{\partial g}{\partial x'} + m\frac{\partial g}{\partial y'} + n\frac{\partial g}{\partial z'}] \sin \theta d\theta d\phi \qquad (4.71)$$

This can be re-expressed as

$$\frac{\partial \bar{g}}{\partial t} = \frac{c}{4\pi c^2 t^2} \int_0^\pi \int_0^{2\pi} [l\frac{\partial g}{\partial x'} + m\frac{\partial g}{\partial y'} + n\frac{\partial g}{\partial z'}] ds \qquad (4.72)$$

Applying the divergence theorem gives

$$\frac{\partial \bar{g}}{\partial t} = \frac{c}{4\pi c^2 t^2} \int \int \int_V (g_{x'x'} + g_{y'y'} + g_{z'z'}) dv \tag{4.73}$$

where the integration is carried over the corresponding volume V of S and dv is the volume element.

Partially differentiating with respect to t again we can write

$$\frac{\partial^2 \bar{g}}{\partial t^2} + \frac{2}{t} \frac{\partial \bar{g}}{\partial t} = \frac{c}{4\pi c^2 t^2} \frac{\partial}{\partial t} \int \int \int_V (g_{x'x'} + g_{y'y'} + g_{z'z'}) dv \tag{4.74}$$

Inserting the specific form of the volume element namely, $dv = r^2 \sin\theta dr d\phi$ where $r = ct$ implies

$$\frac{\partial^2 \bar{g}}{\partial t^2} + \frac{2}{t} \frac{\partial \bar{g}}{\partial t} = \frac{c}{4\pi c^2 t^2} \frac{\partial}{\partial t} \int_0^{ct} \int_0^{\pi} \int_0^{2\pi} (g_{x'x'} + g_{y'y'} + g_{z'z'}) r^2 \sin\theta dr d\theta d\phi \tag{4.75}$$

The partial derivative of t gets rid of the t-integral yielding the simple result

$$\frac{\partial^2 \bar{g}}{\partial t^2} + \frac{2}{t} \frac{\partial \bar{g}}{\partial t} = \frac{c^2}{4\pi} \int_0^{\pi} \int_0^{2\pi} (g_{x'x'} + g_{y'y'} + g_{z'z'}) \sin\theta d\theta d\phi \tag{4.76}$$

Using now the definition (4.68) of ψ which points to

$$\psi_{tt} - c^2 \nabla^2 \psi = (t\bar{g})_{tt} - c^2 \nabla^2 (t\bar{g})$$

gives the form

$$\psi_{tt} - c^2 \nabla^2 \psi = t[(\frac{2}{t} \frac{\partial \bar{g}}{\partial t} + \frac{\partial^2 \bar{g}}{\partial t^2}) - c^2 \nabla^2 \bar{g}]$$

Employing (4.67) and (4.76), the right side of the above can be seen to vanish proving that ψ satisfies the homogeneous wave equation

$$\psi_{tt} - c^2 \nabla^2 \psi = 0 \tag{4.77}$$

We now note the following points:

(i) The continuity of g implies from (4.66) that $\bar{g}$ is continuous too and hence it follows from the definition of ψ that the first criterion of (4.70) holds.

(ii) If we differentiate (4.68) with respect to t we obtain

$$\psi_t = \bar{g} + t\bar{g}_t \tag{4.78}$$

Since the derivative of g is continuous so is $\bar{g}_t$ and is therefore bounded. Hence it follows from (4.78) that the second criterion of (4.70) holds too.

So far we have dealt with the function g. We now focus on f. Let us define a function χ by

$$\chi(x,y,z,t) = t\bar{f}(x,y,z,t) \tag{4.79}$$

where $\bar{f}$ is the spherical mean over S

$$\bar{f}(x,y,z,t) = \frac{1}{4\pi c^2 t^2} \int_S f(x',y',z')ds' \tag{4.80}$$

which, as similar to (4.67), is given by

$$\bar{f} = \frac{1}{4\pi} \int_0^\pi \int_0^{2\pi} f(x+lct, y+mct, z+nct) \sin\theta\, d\theta\, d\phi \tag{4.81}$$

We now show that the following proposition holds:

The function χ is a solution of the homogeneous equation

$$\chi_{tt} - c^2 \nabla^2 \chi(x,y,z,t) = 0 \tag{4.82}$$

obeying the initial conditions

$$\chi(x,y,z,0) = 0, \quad \chi_t(x,y,z,0) = f(x,y,z) \tag{4.83}$$

Proof: To proceed with the proof, we focus on the function χ_t which has continuous derivatives up to second order. Moreover it obeys, on using (4.82)

$$(\chi_t)_{tt} = (\chi_{tt})_t = (c^2 \nabla^2 \chi)_t = c^2 \nabla^2 (\chi_t) \tag{4.84}$$

If for simplicity we set $\xi = \chi_t$ then from the above equation we see that ξ obeys the homogeneous equation

$$\xi_{tt}(x,y,z,t) - c^2 \nabla^2 \xi(x,y,z,t) = 0 \tag{4.85}$$

to be considered along with it the initial condition

$$\xi(x,y,z,0) = f(x,y,z) \tag{4.86}$$

as is clear from the second equation of (4.83).

On the other hand, if we twice differentiate the first equation of (4.83) with respect to the variables x, y and z we obtain

$$\chi_{xx}(x,y,z,0) = 0, \quad \chi_{yy}(x,y,z,0) = 0 \quad \chi_{zz}(x,y,z,0) = 0 \tag{4.87}$$

This has the implication from (4.85) that

$$\xi_t(x,y,z,0) = \chi_{tt}(x,y,z,0) = c^2\nabla^2 = c^2[\chi_{xx} + \chi_{yy} + \chi_{zz}]_{(x,y,z,0)} = 0 \tag{4.88}$$

We therefore find that ξ is a solution of the wave equation (4.85) subject to the initial conditions

$$\xi(x,y,z,0) = f(x,y,z), \quad \xi_t(x,y,z,0) = 0 \tag{4.89}$$

If we now superpose the ψ-equation from (4.77) and ξ-equation from (4.85) and call

$$\Phi = \psi + \xi \tag{4.90}$$

then Φ obeys the wave equation

$$\Phi_{tt}(x,y,z,t) - c^2\nabla^2\Phi(x,y,z,t) = 0 \tag{4.91}$$

To find the accompanying initial conditions to (4.91), we simply have to add (4.70) and (4.89).

To summarize, the initial value problem for the homogeneous form of the PDE

$$\Phi_{tt}(x,y,z,t) - c^2\nabla^2\Phi(x,y,z,t) = 0 \tag{4.92}$$

subject to the inhomogeneous conditions

$$\Phi(x,y,z,0) = f(x,y,z), \quad \Phi_t(x,y,z,0) = g(x,y,z) \tag{4.93}$$

has the following solution for $\Phi(x,y,z,t)$

$$\Phi(x,y,z,t) = \frac{1}{4\pi}\frac{\partial}{\partial t}\int_0^\pi\int_0^{2\pi} tf(x+lct,y+mct,z+nct)\sin\theta d\theta d\phi$$
$$+ \frac{1}{4\pi}\int_0^\pi\int_0^{2\pi} tg(x+lct,y+mct,z+nct)\sin\theta d\theta d\phi \tag{4.94}$$

where l, m, n are defined in (4.63) - (4.65) and we have used (4.68) and (4.79) along with the definition $\xi = \chi_t$. The above integral representation of Φ is called Poisson's formula and gives the propagation of an initial disturbance over the entire space. (4.94) is also called the Kirchhoff's formula and stands for the specific solution of the initial value problem formulated in (4.92) and (4.93). Note that (4.92) is the homogeneous counterpart of the equation (4.60) but tied up here with the set of inhomogeneous initial conditions (4.93).

Step 2

We now proceed to find the solution of the inhomogeneous equation

$$\Psi_{tt}(x, y, z, t) - c^2 \nabla^2 \Psi(x, y, z, t) = \rho(x, y, z, t) \tag{4.95}$$

which is subject to the homogeneous conditions

$$\Psi(x, y, z, 0) = 0, \quad \Psi_t(x, y, z, 0) = 0 \tag{4.96}$$

Suppose t is translationally shifted by a parameter τ and we identify $g(x, y, z, t) \equiv \rho(x, y, z, t - \tau)$. Then by (4.68) ψ given by

$$\psi(x, y, z, t; \tau) = (t - \tau)\bar{\rho}(x, y, z, t - \tau) \tag{4.97}$$

and reads explicitly

$$\psi(x, y, z, t; \tau) = \frac{t - \tau}{4\pi} \int_0^\pi \int_0^{2\pi} \rho(x + lc(t - \tau), y + mc(t - \tau), z + nc(t - \tau); \tau) \sin\theta d\theta d\phi \tag{4.98}$$

and serves as a solution of the wave equation (4.69). The accompanying conditions are

$$\psi(x, y, z, \tau) = 0, \quad \psi_t(x, y, z, \tau) = \rho(x, y, z, \tau) \tag{4.99}$$

where because of the factor in (4.97) the condition $t = 0$ has been transformed to $t = \tau$.

Concerning Ψ we take its form as the integral

$$\Psi(x, y, z, t) = \int_0^t \psi(x, y, z, t; \tau) d\tau \tag{4.100}$$

which implies

$$\Psi(x, y, z, 0) = 0 \tag{4.101}$$

Further partially differentiating both sides of (4.100) with respect to t gives

$$\Psi_t = \psi(x, y, z, t; t) + \int_0^t \psi_t(x, y, z, t; \tau) d\tau \tag{4.102}$$

Because of the vanishing of the first term in the right side of (4.102) due to the first condition of (4.99) at $t = \tau$, Ψ_t corresponds to the integral

$$\Psi_t = \int_0^t \psi_t(x, y, z, t; \tau) d\tau \tag{4.103}$$

and we conclude that

$$\Psi_t(x, y, z, 0) = 0 \tag{4.104}$$

If we partially differentiate again the expression of Ψ_t, we obtain from (4.103)

$$\Psi_{tt} = \psi_t(x, y, z, t; t) + \int_0^t \psi_{tt}(x, y, z, t; \tau) d\tau \tag{4.105}$$

Here the first term in the right side is $\rho(x, y, z, t)$ by the second condition of (4.99) while we can replace ψ_{tt} by $c^2 \nabla^2 \psi$ in the second term. This means that Ψ satisfies the inhomogeneous equation

$$\Psi_{tt} = \rho(x, y, z, t) + c^2 \nabla^2 \int_0^t \psi(x, y, z, t; \tau) d\tau = c^2 \nabla^2 \Psi \tag{4.106}$$

where we have used (4.100). Note that the explicit expression for Ψ, by (4.98) and (4.100), is

$$\Psi(x, y, z, t) = \int_0^t d\tau \frac{t - \tau}{4\pi} \int_0^\pi \int_0^{2\pi} \rho(x + lc(t-\tau), y + mc(t-\tau), z + nc(t-\tau); \tau) \sin\theta d\theta d\phi \tag{4.107}$$

We can set $r = c(t - \tau)$ to represent (4.107) in a more meaningful form from a physical point of view. In fact we find

$$\Psi(x, y, z, t) = \frac{1}{4\pi c^2} \int_0^{ct} \int_0^\pi \int_0^{2\pi} \frac{1}{r} \rho(x + lr, y + mr, z + nr; t - \frac{r}{c}) dv \tag{4.108}$$

where $dv = r^2 \sin\theta d\theta d\phi$. Equivalently, (4.108) can be put as

$$\Psi(x, y, z, t) = \frac{1}{4\pi c^2} \int_V \frac{\rho(x', y', z'; t - \frac{r}{c})}{r} dx' dy' dz' \tag{4.109}$$

where $r = \sqrt{(x - x')^2 + (y - y')^2 + (z - z')^2}$ and quite clearly represents a retarded potential.

We therefore find that while Φ as given by (4.94) is a solution[1] of the problem (4.92) and (4.93), Ψ as given by (4.108) is a solution of the problem (4.95) and (4.96). By applying the principle of superposition $\phi(\equiv \Phi + \Psi)$ is a solution of the problem (4.60) and (4.61).

---

[1] For a qualitative analysis and a discussion of the uniqueness of the problem see *R.B.Guenther and J.W.Lee, Partial differential equations and mathematical physics and integral equations*, Dover Publications, New York (1988), pp. 407–410.

(b) Two dimensional wave equation

The Cauchy problem for the two-dimensional inhomogeneous wave equation is given by the following set of equations

$$\phi_{tt} - c^2(\phi_{xx} + \phi_{yy}) = \rho(x, y, t), \quad -\infty < x, y < +\infty \quad (4.110)$$

where ρ is the inhomogeneous term, subject to the initial conditions

$$\phi(x, y, 0) = f(x, y), \quad \phi_t(x, y, 0) = g(x, y), \quad -\infty < x, y < +\infty \quad (4.111)$$

We require that f has continuous derivatives up to second order and g has continuous derivatives only.

To tackle this problem we follow Hadamard's method of descent wherein we look upon the above problem as a reduced version of the three-dimensional Cauchy problem in the sense that the solution does not depend on the z-variable.

With this aim in mind, let us first focus on the following two-dimensional initial value problem defined in terms of $\psi(x, y, t)$ which satisfies the homogeneous equation

$$\psi_{tt} - c^2 \nabla^2 \psi(x, y, t)- = 0 \quad (4.112)$$

subject to the inhomogeneous initial conditions

$$\psi(x, y, 0) = f(x, y), \quad \psi_t(x, y, 0) = g(x, y) \quad (4.113)$$

Let us recall the Poisson's or Kirchoff's formula for the three-dimensional case we encountered in the previous section. We found for the three-dimensional counterpart of (4.112) the solution (4.94). In terms of ψ it reads

$$\psi(x, y, z, t) = \frac{1}{4\pi} \frac{\partial}{\partial t} \int_0^\pi \int_0^{2\pi} t\alpha(x + lct, y + mct, z + nct) \sin\theta \, d\theta \, d\phi$$

$$+ \frac{1}{4\pi} \int_0^\pi \int_0^{2\pi} t\beta(x + lct, y + mct, z + nct) \sin\theta \, d\theta \, d\phi \quad (4.114)$$

where we call $\psi(x, y, z, 0) = \alpha(x, y, z)$ and $\psi_t(x, y, z, 0) = \beta(x, y, z)$

Note that the above formula could also have been expressed as

$$\psi(x, y, z, t) = \frac{1}{4\pi c^2} \frac{\partial}{\partial t} \int_S \frac{\alpha(x', y', z')}{t} \, ds + \frac{1}{4\pi c^2} \int_S \frac{\beta(x', y', z')}{t} \, ds \quad (4.115)$$

where the integration is over the surface of the sphere S of radius ct with its centre at (x, y, z) and the point (x', y', z') lying anywhere on the surface of S.

If we suppress the dependence on z for ψ and let the functions α and β be independent of the variable z', the above representation reduces to

$$\xi(x,y,t) = \frac{1}{4\pi c^2} \frac{\partial}{\partial t} \int_S \frac{\alpha(x',y')}{t} ds + \frac{1}{4\pi c^2} \int_S \frac{\beta(x',y')}{t} ds \qquad (4.116)$$

where we have replaced ψ by ξ to avoid confusion of notation. Writing f and g in place of α and β respectively it is clear that the following form for $\xi(x,y,t)$

$$\xi(x,y,t) = \frac{1}{4\pi c^2} \frac{\partial}{\partial t} \int_S \frac{f(x',y')}{t} ds + \frac{1}{4\pi c^2} \int_S \frac{g(x',y')}{t} ds \qquad (4.117)$$

is a solution of the problem posed in (4.112) and (4.113).

The solution can be expressed in a different form by considering the projection of S on the $z' = 0$ plane. It is evidently S' which stands for the area of the circle $(x - x')^2 + (y - y')^2 = c^2 t^2$. Now the normal to S at the point (x', y', z') makes an angle θ with the z'-axis given by

$$\cos\theta = \frac{z - z'}{ct} = \frac{1}{ct}\sqrt{c^2 t^2 - (x - x')^2 - (y - y')^2} = \frac{\sqrt{c^2 t^2 - R^2}}{ct} \qquad (4.118)$$

where $R^2 = (x - x')^2 + (y - y')^2$. Therefore the projection ds' of ds on the $z' = 0$-plane is

$$ds' = |\cos\theta| ds = \frac{\sqrt{c^2 t^2 - R^2}}{ct} ds \qquad (4.119)$$

Consequently we can write

$$\int_S \frac{f(x',y')}{t} ds = 2c \int_0^{ct} \int_0^{2\pi} \frac{f(x + R\cos\phi, y + R\sin\phi)}{\sqrt{c^2 t^2 - R^2}} R d\phi dR \qquad (4.120)$$

and

$$\int_S \frac{g(x',y')}{t} ds = 2c \int_0^{ct} \int_0^{2\pi} \frac{g(x + R\cos\phi, y + R\sin\phi)}{\sqrt{c^2 t^2 - R^2}} R d\phi dR \qquad (4.121)$$

where the factor of 2 in the right side appears because of the integration over $\sin\theta$ in the range between 0 and π.

As a result ξ acquires the following form

$$\xi(x,y,t) = \frac{1}{2\pi c} \frac{\partial}{\partial t} \int_0^{ct} \int_0^{2\pi} \frac{f(x + R\cos\theta, y + R\sin\theta)}{\sqrt{c^2 t^2 - R^2}} R d\theta dR$$

$$+ \frac{1}{2\pi c} \int_0^{ct} \int_0^{2\pi} \frac{g(x + R\cos\theta, y + R\sin\theta)}{\sqrt{c^2 t^2 - R^2}} R d\theta dR \qquad (4.122)$$

This above formula is called Poisson-Parseval formula.

Having derived the solution of the two-dimensional homogeneous wave equation with inhomogeneous initial conditions, we attend to the inhomogeneous form

$$\eta_{tt} - c^2(\eta_{xx}(x, y, t) + \eta_{yy}(x, y, t)) = \rho(x, y, t) \tag{4.123}$$

where ρ is the inhomogeneous term, subject to the homogeneous initial conditions

$$\eta(x, y, 0) = 0, \quad \eta_t(x, y, 0) = 0 \tag{4.124}$$

As before taking cue from the three-dimensional case where we found for the inhomogeneous equation

$$\psi_{tt}(x, y, z, t) - c^2(\psi_{xx} + \psi_{yy} + \psi_{zz}) = \rho(x, y, z, t) \tag{4.125}$$

which is subject to the homogeneous conditions

$$\psi(x, y, z, 0) = 0, \quad \psi_t(x, y, z, 0) = 0 \tag{4.126}$$

the solution

$$\psi(x, y, z, t) = \frac{1}{4\pi} \int_0^t \int_0^\pi \int_0^{2\pi} (t-\tau)\rho(x+lc(t-\tau), y+mc(t-\tau), z+nc(t-\tau); \tau) \sin\theta\, d\theta\, d\phi\, d\tau \tag{4.127}$$

where l, m, n have the usual meanings as noted before.

If we now assume ψ and ρ to be independent of the variable z the reduced form of (4.127) in terms of η is

$$\eta(x, y, t) = \frac{1}{4\pi} \int_0^t \int_0^\pi \int_0^{2\pi} (t-\tau)\rho(x+lc(t-\tau), y+mc(t-\tau); \tau) \sin\theta\, d\theta\, d\phi\, d\tau \tag{4.128}$$

and serves as a solution of the problem (4.123) and (4.124). In (4.128) we have replaced ψ by η to avoid confusion of notation.

Notice the symmetry of the integrand in (4.127) under $\theta \to \pi - \theta$ allows us to re-express it as

$$\eta(x, y, t) = \frac{1}{2\pi} \int_0^t \int_0^{\frac{\pi}{2}} \int_0^{2\pi} (t - \tau')\rho(x + lc(t - \tau'), y + mc(t - \tau'); \tau') \sin\theta\, d\theta\, d\phi\, d\tau' \tag{4.129}$$

where $l = \sin\theta\cos\phi$ and $m = \sin\theta\cos\phi$.

Let us now go for the following change of variables

$$X = x + c(t - \tau')\sin\theta\cos\phi, \quad Y = y + mc(t - \tau')\sin\theta\sin\phi, \quad \tau = \tau' \quad (4.130)$$

which means that the region of integration is restricted to the zone

$$(X - x)^2 + (Y - y)^2 \le c^2(t - \tau)^2, \quad 0 \le \tau \le t \quad (4.131)$$

and the Jacobian matrix of the transformation

$$J = \frac{\partial(X, Y, \tau)}{\partial(\theta, \phi, \tau')} \quad (4.132)$$

takes the form

$$J = \begin{pmatrix} \frac{\partial X}{\partial \theta} & \frac{\partial Y}{\partial \theta} & \frac{\partial \tau}{\partial \theta} \\ \frac{\partial X}{\partial \phi} & \frac{\partial Y}{\partial \phi} & \frac{\partial \tau}{\partial \phi} \\ \frac{\partial X}{\partial \tau'} & \frac{\partial Y}{\partial \tau'} & \frac{\partial \tau}{\partial \tau'} \end{pmatrix} \quad (4.133)$$

On performing the partial derivatives we obtain for J

$$J = \begin{pmatrix} c(t - \tau')\cos\theta\cos\phi & c(t - \tau')\cos\theta\sin\phi & 0 \\ -c(t - \tau')\sin\theta\sin\phi & c(t - \tau')\sin\theta\cos\phi & 0 \\ -\sin\theta\cos\phi & -c\sin\theta\sin\phi & 1 \end{pmatrix} \quad (4.134)$$

and as a result the element $dX\,dY\,d\tau$ becomes

$$dX\,dY\,d\tau = c^2(t - \tau')^2\sin\theta\cos\theta\,d\theta\,d\phi\,d\tau' \quad (4.135)$$

It implies

$$(t - \tau')\sin\theta\,d\theta\,d\phi\,d\tau' = \frac{1}{c^2(t - \tau')\cos\theta}dX\,dY\,d\tau = \frac{1}{c\sqrt{c^2(t - \tau)^2 - (X - x)^2 - (Y - y)^2}} \quad (4.136)$$

where (4.130) has been used.

By the above result the expression (4.128) for η becomes

$$\eta(x, y, t) = \frac{1}{2\pi c}\int\int\int_V \frac{\rho(X, Y, \tau)}{\sqrt{c^2(t - \tau)^2 - \epsilon^2}}dX\,dY\,d\tau \quad (4.137)$$

where $\epsilon^2 = (X - x)^2 + (Y - y)^2$ and by the principle of superposition

$$\phi = \xi + \eta \quad (4.138)$$

is a solution of the problem posed in (4.110) and (4.111).

4.5   Summary

The main goal of this chapter was to introduce the wave equation as a typical PDE of hyperbolic type. We started with the D'Alembert solution of the one-dimensional case and inquired into its special properties. Subsequently we discussed the Riemann method of solving the normal form of a hyperbolic equation and the role played by the Riemann function with the characteristic boundary conditions. The Riemann function was found by prescribing the initial data on any smooth non-characteristic curve and the solution was obtained in the form of quadratures. Next, we took up the problem of finding the solution of the wave equation by the method of separation of variables. We considered both the cases of spherical and cylindrical coordinates and wrote down the exact forms of the general nature of the solutions. We then turned to addressing in some detail the initial value problems for the homogeneous and non-homogeneous cases of the three and two-dimensional wave equation and derived Poisson-Kirchoff and Poisson-Parseval formulas.

Exercises

1. Consider the problem of forced vibrations of a string represented by the inhomogeneous equation

$$\phi_{xx}(x, t) = \frac{1}{c^2}\phi_{tt}(x, t) + \rho(x, t)$$

subject to

$$\phi(0, t) = 0 = \phi(l, t)$$

where $x = 0$ and $x = l$ are the end points of the string and the function ρ is assumed to be given. If $\rho(x, t) = \rho(x) \sin \omega t$ then employing the separation of variables $\phi(x, t) = \phi(x) \sin \omega t$ show that the given problem can be converted to the ODE form

$$\phi_{xx}(x) + \frac{\omega^2}{c^2}\phi(x) = \rho(x), \quad \phi(0) = 0 = \phi(l)$$

with the solution $\phi(x) = \lambda_n \sin \frac{n\pi x}{l}$, $\lambda_n \neq 0$ are arbitrary constants.

2. Show that the Cauchy problem for the following hyperbolic equation

$$y^2\phi_{xx} - y\phi_{yy} + \frac{\phi_y}{2} = 0$$

subject to the conditions

$$\phi(x, 0) = f(x), \quad \phi_y(x, 0) = g(x)$$

is not well-posed.

3. Show that the function $\phi(x, y, t) = xyt - \frac{1}{6}xyt^3$ solves the initial value problem of

$$\phi_{xx} + \phi_{yy} - \phi_{tt} = xyt$$

subject to the conditions

$$\phi(x, y, 0) = 0, \quad \phi_t(x, y, 0) = xy$$

4. The equation of an oscillatory string is solved by $\phi(x, t)$ satisfying the equation

$$\phi_{xx} - \phi_{tt} = 0$$

Show that the following function with the specified arguments

$$\phi(\frac{x}{x^2 - t^2}, \frac{t}{x^2 - t^2})$$

also solves the above equation.

5. Consider the Cauchy problem wherein the one-dimensional inhomogeneous wave equation

$$\phi_{tt} = c^2 \phi_{xx} + \rho(x,t), \quad -\infty < x < \infty, \quad t > 0$$

is guided by the initial conditions

$$\phi(x,0) = 0 = \phi_t(x,0), \quad -\infty < x < \infty$$

Show that if $\phi(x,t)$ is an odd function with respect to the variable x then $\phi(x,t)$ is vanishing at $x = 0$ while if $\phi(x,t)$ is an even function with respect to the variable x then the spatial-derivative $\phi_x(x,t)$ is vanishing at $x = 0$.

6. Suppose that we are given a second order operator

$$L \equiv A\frac{\partial^2}{\partial x^2} + 2B\frac{\partial^2}{\partial x \partial y} + C\frac{\partial^2}{\partial y^2}$$

where A, B, C are constant coefficients. By factorizing L in terms of first order entities ξ_1 and ξ_2 as given by

$$\xi_i = a_i\frac{\partial}{\partial x} + b_i\frac{\partial}{\partial y} + c_i, \quad i = 1,2$$

show that a general solution to $L\phi = 0$ can be written as the superposition $\phi = \phi_1 + \phi_2$, where ϕ_1 and ϕ_2 are solutions of $\xi_1\phi_1 = 0$ and $\xi_2\phi_2 = 0$.
  Hence establish that for the PDE

$$\frac{\partial^2 \phi}{\partial x \partial y} + 2\frac{\partial^2 \phi}{\partial y^2} - \frac{\partial \phi}{\partial x} - 2\frac{\partial \phi}{\partial y} = 0$$

its general solution is expressible as

$$\phi(x,y) = f(2x - y) + e^y g(x)$$

for arbitrary functions f and g.

7. Show that for the one-dimensional Klein-Gordon equation

$$\phi_{tt} - c^2 \phi_{xx} + k^2 \phi = 0$$

where k^2 is a constant, the Riemann function is of the form

$$\chi(x,y;\xi,\eta) = J_0[ik\sqrt{(x-\xi)(y-\eta)}]$$

where J_0 is the Bessel function of order zero.

8. Show for the Euler-Poisson equation

$$\phi_{xy} + \frac{k}{x+1}(\phi_x + \phi_y) = 0$$

where k is a constant, the Riemann function is of the form

$$\chi(x, y; \xi, \eta) = (\frac{x+y}{\xi+\eta})^k F(1 - k, k, 1; \frac{(\xi - x)(\eta - y)}{(\xi + \eta)(x + y)})$$

where F is the hypergeometric function.

9. Given the inhomogeneous PDE

$$\phi_{tt} = c^2 \phi_{xx} + \cos(x)$$

with the initial conditions

$$\phi(x, 0) = \sin(x), \quad \phi_t(x, 0) = 1 + x$$

find the following solution

$$\phi(x, t) = \sin(x)\cos(ct) + xt + t + \frac{1}{c^2}(\cos(x) - \cos(x)\cos(ct))$$

10. Consider the inhomogeneous hyperbolic equation in $1 + 1$ dimensions

$$\phi_{tt} = \phi_{xx} + \sin(\omega t)\sin(x)$$

Depending on whether the frequency ω is unity or not show that that the solution assumes two different forms

$$\phi(x, t) = \begin{cases} \frac{\sin(\omega t) - \omega \sin(t)}{1 - \omega^2} \sin(x), & 0 < \omega \neq 1 \\ \frac{\sin(t) - t\cos(t)}{2} \sin(x), & \omega = 1 \end{cases} \tag{4.139}$$

Note that the case $\omega = 1$ reveals a bounded nature of the solution whereas for $\omega = 1$ the solution blows up asymptotically with respect to t. This case corresponds to ω being a resonant frequency.

11. The nonlinear hyperbolic sine-Gordon equation in $1+1$ dimensions is given by

$$\phi_{tt} - \phi_{xx} + \sin\phi = 0$$

Its solutions are defined on the whole real line and possess the decay property $\lim_{|x| -> \infty} \phi(x, t) = 0$. Obtain by substitution the following solution

$$\phi(x, t) = 4 \tan^{-1}\left[\frac{\sqrt{1 - \omega^2} \cos(\omega t)}{\omega \cosh(\sqrt{1 - \omega^2}x)}\right]$$

Such a solution is referred to as the breather.[2] Notice that it is time-periodic and localized in space. The sine-Gordon equation is the sine version of the well known Klein-Gordon equation.

---

[2]The above is a 1-soliton solution. Sine-Gordon equation enjoys multi-soliton solutions too. The Lagrangian for the sine-Gordon equation is given by $L = \frac{1}{2}(\phi_t^2 - \phi_x^2) + \cos \phi - 1$.

Chapter 5

PDE: Parabolic form

5.1   Reaction-diffusion and heat equations

Reaction-diffusion equation

We start with a brief derivation of the reaction-diffusion equation. Let a uniform circular tube of cross-sectional area A be held in such a way that that the x-axis is chosen to coincide with its axial direction. We focus on the portion of the tube marked with $x = a$ and $x = b$ (see Figure 5.1). We consider the movement of some substance in this tube. Let the substance be governed by the density function $\phi(x, t)$, x and t denoting respectively the location and time. The total amount of the substance inside the tube would be

$$\Phi(t) = \int_a^b \phi(x, t) A dx \qquad (5.1)$$

Typically, the substance can correspond to a population in which case $\phi(x, t)$ stands for the population density. But here we will be interested with a flow possessing properties like continuity and differentiability up to sufficient orders.

To inquire into the change of Φ with evolution of time, we note that two things can contribute to it. One is that there is a continuous creation or destruction of the substance due to certain physical reasons that we attribute to the presence of some some external or internal agency. Specifically, if there is some source function f as defined by

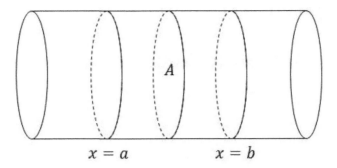

FIGURE 5.1: A tube of cross-sectional area A.

$$f > 0 : \text{source} \tag{5.2}$$

it would give the rate r of generation of the substance in the tube

$$r = \int_a^b f A dx \tag{5.3}$$

where, in principle, $f = f(x, t, \phi)$.

The other factor influencing the motion of the flow is the flux which denotes the amount of the substance that passes through the area at the point x at time t. If $\xi(x, t)$ define the flux function, then the rate of flux across the boundary is

$$q = A[\xi(a, t) - \xi(b, t)] \tag{5.4}$$

By the principle of conservation we have from (5.1), (5.3) and (5.4) the balance law

$$\frac{d\Phi}{dt} = q + r \tag{5.5}$$

which implies

$$\int_a^b [\phi_t(x, t) + \xi_x(x, t) - f(x, t, \phi)] dx = 0 \tag{5.6}$$

The continuity of the integrand allows us to conclude

$$\phi_t(x, t) + \xi_x(x, t) - f(x, t, \phi) = 0, \quad t > 0 \tag{5.7}$$

About $\xi(x,t)$, Fick's law makes an educated approximation that it is proportional to the gradient of the density as is often found to be true in many physical processes. Thus

$$\xi(x,t) = -\kappa\phi_x(x,t) \tag{5.8}$$

where κ is a diffusion coefficient which can be a position-dependent quantity. Its dimension is the ratio of the square of the length and time. We are therefore led to the form

$$\phi_t(x,t) - (\kappa\phi_x)_x = f(x,t,\phi), \quad t > 0 \tag{5.9}$$

(5.9) is called the reaction-diffusion equation.

In the absence of a source term we have the reduced version of the diffusion equation

$$\phi_t(x,t) - (\kappa\phi_x)_x = 0, \quad t > 0 \tag{5.10}$$

An appropriate boundary condition can be imposed if an insulation condition is assumed at the end point $x = 0$ i.e. there is no flux at $x = 0$:

$$\phi_x(0,t) = 0, \quad x = 0 \tag{5.11}$$

Note that the two or three-dimensional analog of (5.10) is

$$\phi_t(x,t) - \nabla \times (\kappa\nabla\phi) = 0, \quad t > 0 \tag{5.12}$$

where ∇ is a two or three-dimensional vector differential operator.

Heat conduction equation

In the heat equation κ is treated as a constant. The one-dimensional heat conduction equation has the form

$$\phi_t(x,t) = \alpha^2\phi_{xx}(x,t), \quad |x| < \infty, \quad t > 0 \tag{5.13}$$

where $\alpha^2 = \kappa$.

In two or three dimensions the heat conduction equation has the form

$$\phi_t(\vec{r},t) = \alpha^2\nabla^2\phi(\vec{r},t), \quad t > 0 \tag{5.14}$$

5.2   Cauchy problem: Uniqueness of solution

In this section we attempt to establish the uniqueness of the solution for the one-dimensional heat equation with respect to the Cauchy problem. It states that if $\phi(x,t)$ is a solution of the heat equation (5.13) obeying

$$\left| \int_{-\infty}^{+\infty} \phi(x,t)dx \right| < \infty \tag{5.15}$$

along with fulfilling the conditions

$$\phi(x,0) = f(x), \quad x \in \Re, \quad \phi_x(x,t) \to 0 \quad \text{as} \quad x \to \pm\infty \tag{5.16}$$

then the heat equation has a unique solution.

Proof: The proof is rather simple and goes as follows. We assume, if possible, two plausible solutions of the heat equation and name them as $\phi_1(x,t)$ and $\phi_2(x,t)$. If $\psi(x,t)$ is the difference of the two solutions then it satisfies the boundary value problem

$$\psi_t(x,t) = a^2 \psi_{xx}(x,t), \quad t > 0 \tag{5.17}$$

subject to

$$\psi(x,0) = 0, \quad x \in \Re, \quad \psi_x(x,t) \to 0 \quad \text{as} \quad x \to \pm\infty, \quad t > 0 \tag{5.18}$$

Define a time-dependent quantity $\Psi(t)$ given by the integral

$$\Psi(t) = \int_{-\infty}^{+\infty} \psi^2(x,t)dx \ge 0, \quad \Psi(0) = 0 \tag{5.19}$$

Taking the time-derivative of both sides results in

$$\frac{d\Psi}{dt} = 2\int_{-\infty}^{+\infty} \psi\psi_t dx = 2a^2 \int_{-\infty}^{+\infty} \psi\psi_{xx} dx \tag{5.20}$$

Integrating by parts and noting that $\psi_x(x,t) \to 0$ as $x \to \pm\infty$ immediately gives the inequality

$$\frac{d\Psi}{dt} = -2a^2 \int_{-\infty}^{+\infty} \psi_x^2 dx \le 0 \tag{5.21}$$

It follows that $\Psi(t)$ is a non-increasing function of t to be compared with the claim in (5.19) that $\Psi(t) \ge 0$ and $\Psi(0) = 0$. The only way of reconcilement is $\Psi(t) = 0$ which implies that $\psi = 0$. In other words, $\phi_1(x,t) = \phi_2(x,t)$ and we have proved the uniqueness criterion.

5.3 Maximum-minimum principle

We focus on the one-dimensional inhomogeneous heat equation defined over a space-time rectangular domain $\Omega : [0, l] \times [0, \tau]$

$$\phi_t(x, t) - \alpha^2 \phi_{xx}(x, t) = \rho(x, t), \quad 0 \le x \le l, \quad 0 \le t \le \tau \qquad (5.22)$$

where ρ is the inhomogeneous term and $\phi(x, t)$ is continuous in the region Ω. We are going to show that if $\rho < 0$ then the maximum of $\phi(x, t)$ is attained either at $t = 0$ or on one of the boundaries $x = 0$ or $x = l$. However, the maximum is never attained in the interior of Ω or at $t = \tau$. On the other hand, if $\rho > 0$ then the minimum of $\phi(x, t)$ is attained either at $t = 0$ or on one of the boundaries $x = 0$ or $x = l$. However, the minimum is never attained in the interior of Ω or at $t = \tau$.

Proof: We prove by contradiction. First we take $\rho < 0$. Then if the maximum of ϕ occurs at an interior point of Ω then at this point $\phi_t = 0, \phi_x = 0$ and $\phi_{xx} \le 0$ and we run into a contradiction by looking for consistency with the inhomogeneous equation (5.20). However, if the maximum of ϕ occurs at $t = \tau$, then at this point which is on the boundary of Ω, ϕ could be increasing and we have $\phi_t \ge 0, \phi_x = 0$ and $\phi_{xx} \le 0$. Again we run into a contradiction on looking for consistency with the inhomogeneous equation. Hence we conclude that the maximum must occur at $t = 0$ or on one of the boundaries $x = 0$ or $x = l$ and not in the interior.

The lines of arguments for the case $\rho > 0$ are exactly similar and we arrive at the conclusion that the minimum must occur at $t = 0$ or on one of the boundaries $x = 0$ or $x = l$ and not in the interior.

We now turn to the homogeneous portion

$$\phi_t(x, t) - \alpha^2 \phi_{xx}(x, t) = 0, \quad 0 \le x \le l, \quad 0 \le t \le \tau \qquad (5.23)$$

where $\phi(x, t)$ is continuous in the region Ω. We will presently see that the maximum and minimum of $\phi(x, t)$ are attained at $t = 0$ or on one of the points on the boundary $x = 0$ or $x = l$. As with the inhomogeneous case here too the maximum or minimum is never attained in the interior of Ω or at $t = \tau$.

Proof: We first address the maximum case. Let us follow the simplest approach by defining a function $\chi(x, t)$

$$\chi(x, t) = \phi(x, t) + \epsilon x^2, \quad \epsilon > 0 \qquad (5.24)$$

where $\phi(x, t)$ solves the homogeneous equation (5.23). Since $\phi(x, t)$ is continuous in Ω, $\chi(x, t)$ too is continuous in Ω and attains a maximum in Ω, say, at $(\bar{x}.\bar{t})$ in Ω. Further, since $\phi(x, t)$ is a solution of the homogeneous equation we observe that

$$\chi_t(x, t) - \alpha^2 \chi_{xx}(x, t) = \phi_t(x, t) - \alpha^2 \phi_{xx}(x, t) - 2\alpha^2 \epsilon = -2\alpha^2 \epsilon < 0 \qquad (5.25)$$

We thus find that $\chi(x,t)$ solves an inhomogeneous equation.

If $(\bar{x}.\bar{t})$ is an interior point of Ω then in it

$$\chi_t(\bar{x}.\bar{t}) \geq 0, \quad \chi_{xx}(\bar{x}.\bar{t}) \leq 0, \tag{5.26}$$

The two inequalities in (5.26) imply $\chi_t(x,t) - \alpha^2 \chi_{xx}(x,t) \geq 0$ which is to be compared with the validity of (5.25). Taken together it follows that

$$0 \leq \chi_t(x,t) - \alpha^2 \chi_{xx}(x,t) < 0 \tag{5.27}$$

which is clearly a contradiction. Since $\chi(x,t)$ is a solution of an inhomogeneous equation we can use the earlier result to claim that $\chi(x,t)$ attains its maximum value M on the initial line $t = 0$ or on the boundary $x = 0$ or $x = l$.

From the definition of $\chi(x,t)$ we can write

$$\chi(x,t) = \phi(x,t) + \epsilon x^2 \leq M + \epsilon l^2, \quad 0 \leq x \leq l, \quad 0 \leq t \leq \tau \tag{5.28}$$

and as a consequence obtain

$$\chi(x,t) \leq M, \quad 0 \leq x \leq l, \quad 0 \leq t \leq \tau \tag{5.29}$$

since ϵ is an arbitrary parameter. Thus the maximum principle holds for the homogeneous heat equation.

The result for the minimum case follows from the criterion that if $\phi(x,t)$ is a solution of the homogeneous heat equation then so is $-\phi(x,t)$. Since the maximum principle holds for $\phi(x,t)$, it follows that if m is the minimum of $\phi(x,t)$, then $-m$ is the maximum value of $-\phi(x,t)$ in Ω and so our conclusion would be that $\phi(x,t)$ assumes its minimum on $t = 0$ or on the boundary $x = 0$ or $x = l$. In other words we have the result

$$m \leq \chi(x,t) \leq M, \quad 0 \leq x \leq l, \quad 0 \leq t \leq \tau \tag{5.30}$$

Example 5.1

Make use of the maximum-minimum principle to estimate the maximum and minimum of the function $f(x,t) = 1 - x^2 - 2\alpha^2 t(1 + 3\nu x) - \nu x^3$ where $\nu > 0$ and $x \in [0,1]$ in the context of the heat equation $\phi_t = \alpha^2 \phi_{xx}$, $0 \leq t \leq T$, T is fixed.

We first note that since $f_{xx} = -2 - 6\nu x$ and $f_t = -6\nu\alpha^2 x - 2\alpha^2$, $f(x,t)$ is a solution of the heat equation. Also, the given function $f(x,t) < 1$ unless $x = 0, t = 0$ in which case it is equal to 1. Hence, $f(x,t) \leq 1$ and 1 is the maximum value of $f(x,t)$ at $(x = 0, t = 0)$. On the other hand, in the given intervals of x and t, $f(x,t) \geq -2\alpha^2 T(1 + 3\nu) - \nu$ indicating that $-6\nu\alpha^2 T - 2\alpha^2 T - \nu$ is the minimum value of $f(x,t)$ at $(x = 1, t = T)$.

The above estimates of the maximum and minimum can be verified from the maximum-minimum principle of the heat equation. We know that, given the whole rectangle $[0, 1] \times [0, T]$, the maximum or minimum is attained either on $t = 0$ or on one of the other sides $x = 0$ or $x = 1$. Specifically, we see that at $t = 0$, $f(x, 0) = 1 - x^2 - \nu x^3$ which clearly shows its maximum is 1 at $x = 0$ and minimum is $-\nu$ at $x = 1$. For the side where $x = 0$, $f(0, t) = 1 - 2\alpha^2 t$, maximum or minimum occurs respectively at 1 when $t = 0$ and $1 - 2\alpha^2 T$ where $t = T$ while for the side $x = 1$ for which $f(1, t) = -6\nu\alpha^2 t - 2\alpha^2 t - \nu$, maximum or minimum occurs respectively at $-\nu$ when $t = 0$ and $-6\nu\alpha^2 T - 2\alpha^2 T - \nu$ at $t = T$. We therefore conclude that the maximum value of 1 for f occurs at $(x = 0, t = 0)$ and the minimum value of $-6\nu\alpha^2 T - 2\alpha^2 T - \nu$ for f occurs at $(x = 1, t = T)$ thereby justifying the maximum-minimum principle of the heat equation.

5.4   Method of separation of variables

(a) Cartesian coordinates (x, y, z)

We focus on the three-dimensional Cartesian coordinates (x, y, z) in terms of which the three-dimensional heat conduction equation (5.14) reads

$$\phi_{xx} + \phi_{yy} + \phi_{zz} = \alpha^2 \phi_t \tag{5.31}$$

where $\phi = \phi(x, y, z, t)$. We seek solutions in the variable-separated product form namely,

$$\phi(x, y, z, t) = F(x)G(y)H(z)T(t) \tag{5.32}$$

This gives (5.31) the transformed ODE

$$\frac{1}{F}\frac{d^2 F}{dx^2} + \frac{1}{G}\frac{d^2 G}{dy^2} + \frac{1}{H}\frac{d^2 H}{dz^2} = \frac{\alpha^2}{T}\frac{dT}{dt} \tag{5.33}$$

Since the left-side of (5.33) is a function of the coordinates x, y and z and the right side is a function of t, consistency requires that each side must be a constant which we take as $-k^2$. The t-integration at once separates out yielding the solution

$$T(t) = \lambda e^{-\frac{k^2}{\alpha^2} t} \tag{5.34}$$

and the remaining part reads

$$\frac{1}{F}\frac{d^2 F}{dx^2} + \frac{1}{G}\frac{d^2 G}{dy^2} + \frac{1}{H}\frac{d^2 H}{dz^2} = -k^2 \tag{5.35}$$

The left side of the above equation has three second order ordinary derivatives in x, y and z respectively. Taking any one to the right side, say, the z-dependent piece, gives the right side in a variable-separated form with respect to the left side. Equating it to a constant, say $-r^2$ we get

$$\frac{d^2 H}{dz^2} + s^2 H = 0 \tag{5.36}$$

where $s^2 = k^2 - r^2$.

Carrying on the separation process with respect to the remaining two variables x and y yields two more ODEs namely,

$$\frac{1}{F}\frac{d^2 F}{dx^2} = -\frac{1}{G}\frac{d^2 G}{dy^2} - r^2 = -a^2 \tag{5.37}$$

where $-a^2$ is a separation constant, This leads to the pair of equations

$$\frac{d^2 F}{dx^2} + a^2 F = 0, \quad \frac{d^2 G}{dy^2} + b^2 G = 0 \tag{5.38}$$

where $b^2 = r^2 - a^2$.

Since the differential equation (5.36) for the function H and the ones in (5.38) for F and G have solutions given by a combination of sine and cosine functions, the general solution $\phi(x, y, z, t)$, which includes the solution (5.34) of the T-equation, emerges as

$$\phi(x, y, z, t) = (c_1 \cos ax + c_2 \sin ax)(c_3 \cos by + c_4 \sin by)(c_5 \cos sz + c_6 \sin sz)e^{-\frac{k^2}{\alpha^2}t} \tag{5.39}$$

where $k^2 = a^2 + b^2 + s^2$, $c_1, c_2, c_3, c_4, c_5, c_6$ are constants and the constant λ has been absorbed in one of the coefficients.

The procedure to handle the one-dimensional and two-dimensional cases is much simpler and is illustrated in the two worked-out problems below.

Example 5.2

The temperature $\phi(x, t)$ in a rod of length l i.e. $0 \le x \le l$ is initially given by the constant ϕ_0 and the ends of the rod are kept at zero temperature

$$\phi(0, t) = \phi(l, t) = 0$$

Obtain the temperature profile of the rod at time t and at position x.

We assume that $\phi(x, t)$ can be expressed in a separation of variable form :

$$\phi(x, t) = X(x)T(t)$$

Then the one-dimensional heat conduction equation (5.13) gives

$$\frac{X''}{X} = \frac{T'}{\alpha^2 T} = k \text{ (constant)}$$

Solving for X reveals the following possibilities

$$X(x) = \begin{cases} \lambda \, \exp(\sqrt{k}x) + \mu \, \exp(-\sqrt{k}x) & k > 0 \\ \lambda x + \mu & k = 0 \\ \lambda \, \cos(\sqrt{-k}x) + \mu \, \sin(\sqrt{-k}x) & k < 0 \end{cases}$$

where λ and μ are constants.

Since we have to meet the boundary conditions $X(0) = X(l) = 0$, only the sine part of the third solution is relevant :

$$X(x) = \mu \, \sin(\frac{n\pi x}{l})$$

where $\sqrt{-k} = \frac{n\pi}{l}$.

Employing now the solution of the T-equation which reads $T'(t) = k\alpha^2 T$ we deduce for $\phi(x,t)$ the form

$$\phi(x,t) = \sum_{n=1}^{\infty} \mu_n \, \sin(\frac{n\pi x}{l}) \exp[-\frac{n^2\pi^2\alpha^2 t}{l^2}]$$

on using the above value of $k = -\frac{n^2\pi^2}{l}$.

To meet the condition $\phi(x,0) = \phi_0$, we find from the above representation

$$\phi_0 = \sum_{n=1}^{\infty} \mu_n \, \sin(\frac{n\pi x}{l})$$

Inverting gives

$$\begin{aligned} \mu_n &= \frac{2\phi_0}{l} \int_0^l \sin\frac{n\pi x}{l} \, dx = \frac{2\phi_0}{l} \frac{l}{n\pi} [\cos(\frac{n\pi x}{l})]_0^l \\ &= \frac{2\phi_0}{n\pi} [(-1)^n - 1] \\ &= \begin{cases} -\frac{4\phi_0}{n\pi} & n \text{ odd} \\ 0 & n \text{ even} \end{cases} \end{aligned}$$

Hence the required form for $\phi(x,t)$ is

$$\begin{aligned} \phi(x,t) &= -\frac{4\phi_0}{\pi} \sum_{n=\text{odd}} \frac{1}{n} \sin(\frac{n\pi x}{l}) \exp[-\frac{n^2\pi^2\alpha^2 t}{l^2}] \\ &\to 0 \text{ as } t \to \infty \end{aligned}$$

at all points of the rod.

Example 5.3

Solve the problem of thermal waves in half space ($z > 0$) wherein the temperature distribution obeys the PDE

$$\phi_t(z,t) = \alpha^2 \phi_{zz}(z,t), \quad t > 0$$

subject to the boundary condition at $z = 0$

$$\phi(0,t) = \phi_0 \cos \omega t$$

where the frequency ω is assumed to be known.

To obtain the solution we apply the process of separation of variables and assume that it is of oscillatory type

$$\phi(z,t) = Re\phi(z)e^{-i\omega t}$$

at the applied frequency ω. Note that at $z = 0$ the given condition is

$$\phi(0,t) = Re\phi_0 e^{-i\omega t}$$

with the complex amplitude $\phi(z)$ satisfying the ODE

$$\frac{d^2\phi}{dz^2} = -\frac{i\omega}{\alpha^2}\phi$$

Its solution can be put in the form

$$\phi(z) = Ae^{kz}$$

where A is an overall constant and k^2 is constrained by

$$k^2 = \frac{\omega}{\alpha^2}e^{-\frac{i\pi}{2}} \quad \rightarrow \quad k_\pm = \pm\sqrt{\frac{\omega}{\alpha^2}}e^{-i\frac{\pi}{4}} = \pm(\frac{\omega}{2\alpha^2})^{\frac{1}{2}}(1-i)$$

The solution corresponding to the positive sign blows up as $z \to \infty$ and so is to be discarded. The acceptable form of the solution is then

$$\phi(z) = A\exp[-(\frac{\omega}{2\alpha^2})^{\frac{1}{2}}(1-i)z] = A\exp(\frac{iz}{d} - \frac{z}{d})$$

where

$$d = (\frac{2\alpha^2}{\omega})^{\frac{1}{2}} = (\frac{\tau\alpha^2}{\pi})^{\frac{1}{2}}$$

stands for the penetration depth corresponding to the frequency ω and period $\tau = \frac{2\pi}{\omega}$.

Identifying $A = \phi_0$ gives the desired solution

$$\phi(z,t) = Re\phi(z)e^{-i\omega t} = \phi_0 e^{-\frac{z}{d}}\cos(\frac{z}{d} - \omega t)$$

indicating an exponentially damped nature of the temperature profile.

Example 5.4

In the context of an anisotropic diffusivity solve the heat equation

$$\phi_t(x, y, t) = \alpha^2 \phi_{xx}(x, y, t) + \beta^2 \phi_{yy}(x, y, t), \quad \alpha \neq \beta, \quad t > 0$$

on a rectangular region defined by $x \in (0, a)$ and $y \in (0, b)$. The given boundary conditions are

$$\phi(0, y, t) = 0, \quad \phi_y(x, 0, t) = 0, \quad \phi(a, y, t) = 0, \quad \phi_y(x, b, t) = 0$$

while the initial condition is

$$\phi(x, y, 0) = s(x, y)$$

To obtain the solution we apply the process of separation of variables in the following product form

$$\phi(x, y, t) = F(x)G(y)T(t)$$

This implies that the boundary conditions are converted to

$$F(0) = 0, \quad F(a) = 0, \quad G'(0) = 0, \quad G'(b) = 0$$

We now substitute our chosen form for $\phi(x, y, t)$ in the given equation. Taking the separation constant as $-k$ we readily obtain the set of ODEs

$$\alpha^2 \frac{F''(x)}{F(x)} + \beta^2 \frac{G''(y)}{G(y)} = \frac{T''(t)}{T(t)} = -k$$

where the primes denote the corresponding derivatives.

Solving for the T-equation we find

$$T(t) = \lambda e^{-kt}$$

where λ is an overall constant.

The F-equation has the form

$$F''(x) + \frac{q}{\alpha^2} F(x) = 0, \quad F(0) = 0, \quad F(l) = 0$$

where q is another separation constant. Its solution that fits the boundary conditions is

$$F_n(x) = \sin(\frac{m\pi x}{a}), \quad m = 1, 2, \dots$$

Next, the G-equation has the form

$$G''(y) + \frac{\kappa}{\beta^2} G(y) = 0, \quad G'(0) = 0, \quad G'(b) = 0$$

where κ is given by

$$\kappa = k - \alpha^2 \left(\frac{n\pi}{a}\right)^2, \quad n = 1, 2, \ldots$$

The solution which fits the derivative boundary conditions of G at $y = 0$ and $y = b$ is

$$G_n(y) = \cos\left(\frac{n\pi y}{b}\right), \quad n = 1, 2, \ldots$$

Thus for a specific pair of indices m and n, $\phi(x, y, t)$ assumes the form

$$\phi(x, y, t) = \sin\left(\frac{m\pi x}{a}\right) \cos\left(\frac{n\pi y}{b}\right) e^{-k_{mn} t}, \quad m, n = 1, 2, \ldots$$

where k_{mn} is given by

$$k_{mn} = \pi^2 \left(\alpha^2 \frac{m^2}{a^2} + \beta^2 \frac{n^2}{b^2}\right)$$

The general solution is the result of superposition over all m and n

$$\phi(x, y, t) = \sum_{m=1}^{\infty} \sum_{n=1}^{\infty} \xi_{mn} \sin\left(\frac{m\pi x}{a}\right) \cos\left(\frac{n\pi y}{b}\right) e^{-k_{mn} t}$$

where ξ_{mn} is a set of unknown coefficients. The latter can be fixed on using the initial condition at $t = 0$

$$s(x, y) = \sum_{m=1}^{\infty} \sum_{n=1}^{\infty} \xi_{mn} \sin\left(\frac{m\pi x}{a}\right) \cos\left(\frac{n\pi y}{b}\right)$$

To invert the above expression for the determination of ξ_{mn} we need to multiply both sides by the product $\sin\left(\frac{l\pi x}{a}\right) \cos\left(\frac{l'\pi y}{b}\right)$ and integrate over the whole rectangle

$$\int_0^a \int_0^b dx\, dy\, s(x, y) \sin\left(\frac{l\pi x}{a}\right) \cos\left(\frac{l'\pi y}{b}\right)$$
$$= \sum_{m=1}^{\infty} \sum_{n=1}^{\infty} \xi_{mn} \int_0^a dx \sin\left(\frac{m\pi x}{a}\right) \sin\left(\frac{l\pi x}{a}\right) \int_0^b dy \cos\left(\frac{n\pi y}{b}\right) \cos\left(\frac{l'\pi y}{b}\right)$$

Since the two integrals in the right side are respectively equal to $\frac{a}{2}\delta_{ml}$ and $\frac{b}{2}\delta_{nl'}$, from the orthogonality conditions of sine and cosine function, the entire right side has the value $\xi_{ll'} \frac{ab}{4}$. Hence we arrive at the following result for ξ_{mn}

$$\xi_{mn} = \frac{4}{ab} \int_0^a \int_0^b dx\, dy\, s(x, y) \sin\left(\frac{m\pi x}{a}\right) \cos\left(\frac{n\pi y}{b}\right)$$

(b) Three dimensions: spherical polar coordinates (r, θ, ϕ)

We will now be concerned with seeking solutions of the three-dimensional heat conduction equation

$$\phi_t(\vec{r}, t) = \alpha^2 \nabla^2 \phi(\vec{r}, t), \quad t > 0 \tag{5.40}$$

in spherical polar coordinates. To this end we assume that the following decomposition holds

$$\phi(\vec{r}, t) = R(r)F(\theta)G(\phi)T(t) \tag{5.41}$$

Here $\vec{r} \equiv (r, \theta, \phi)$ are the spherical polar coordinates and ∇^2 has the known representation

$$\nabla^2 = \frac{1}{r^2} \frac{\partial}{\partial r} \left(r^2 \frac{\partial}{\partial r} \right) + \frac{1}{r^2 \sin \theta} \frac{\partial}{\partial \theta} \left(\sin \theta \frac{\partial}{\partial \theta} \right) + \frac{1}{r^2 \sin^2 \theta} \frac{\partial^2}{\partial \phi^2} \tag{5.42}$$

as already furnished in (3.27).

Substitution of the ∇^2 in the heat equation facilitates conversion to the following ODE

$$\frac{1}{R} \frac{d}{dr} \left(r^2 \frac{dR}{dr} \right) + \frac{1}{F} \frac{1}{r^2 \sin \theta} \frac{d}{d\theta} \left(\sin \theta \frac{dF}{d\theta} \right) + \frac{1}{r^2 \sin^2 \theta} \frac{1}{G} \frac{d^2 G}{d\phi^2} = \frac{\alpha^2}{T} \frac{dT}{dt} \tag{5.43}$$

Since the left-side of (5.43) is a function of r, θ and ϕ and the right side is a function of t, consistency requires that each side must be a constant which we take as $-k^2$. The t-integration at once separates out and provides the solution

$$T(t) = \lambda e^{-\frac{k^2}{\alpha^2} t} \tag{5.44}$$

The remaining part has the form

$$r^2 \sin^2 \theta \left[\frac{1}{R} \frac{d}{dr} \left(r^2 \frac{dR}{dr} \right) + \frac{1}{F} \frac{1}{r^2 \sin \theta} \frac{d}{d\theta} \left(\sin \theta \frac{dF}{d\theta} \right) + k^2 \right] = -\frac{1}{G} \frac{d^2 G}{d\phi^2} = m^2 \tag{5.45}$$

where we have put the constant m^2 to signal that the variable ϕ has been separated out.

The G-equation reads explicitly

$$\frac{d^2 G}{d\phi^2} + m^2 G = 0 \tag{5.46}$$

whose general solution is

$$G(\phi) = a \cos(m\phi) + b \sin(m\phi) \tag{5.47}$$

where a and b are constants. There is a restriction on m to be $m = 0.1.2., , ,$ because for a physical system m and $m + 2\pi$ need to represent the same point.

Turning to r and θ part we see that they can be readily disentangled giving

$$\frac{1}{R}\frac{d}{dr}(r^2\frac{dR}{dr}) + k^2r^2 = \frac{m^2}{\sin^2\theta} - \frac{1}{F\sin\theta}\frac{d}{d\theta}(\sin\theta\frac{dF}{d\theta}) = n(n+1) \qquad (5.48)$$

where to acquire a standard representation we have put the separation constant as $n(n+1)$. As a result the R-equation takes the form

$$\frac{d^2R}{dr^2} + \frac{2}{r}\frac{dR}{dr} + [k^2 - \frac{n(n+1)}{r^2}]R = 0 \qquad (5.49)$$

while the F-equation becomes

$$\frac{d^2F}{d\theta^2} + \cot\theta\frac{dF}{d\theta} + [n(n+1) - \frac{m^2}{\sin^2\theta}]F = 0 \qquad (5.50)$$

One can recognize the R-equation to be a spherical Bessel equation and its solutions are called spherical Bessel functions. We write the solution $R(r)$ as

$$R(r) = Aj_n(kr) + By_n(kr) \qquad (5.51)$$

where A and B are constants and the links of $j_n(kr)$ and $y_n(kr)$ to Bessel function J_n and Neumann function Y_n are respectively given by

$$j_n(kr) = \sqrt{\frac{\pi}{2r}}J_{n+\frac{1}{2}}(r), \quad y_n(kr) = \sqrt{\frac{\pi}{2r}}Y_{n+\frac{1}{2}}(r) \qquad (5.52)$$

Note that in the steady state system when k vanishes, $R(r)$ reduces to the simple form

$$R(r) = Cr^n + Dr^{-n-1} \qquad (5.53)$$

where C and D are constants.

On the other hand, the F equation depicts an associated Legendre equation and its solution can be expressed in terms of associated Legendre functions

$$F(\theta) = EP_n^m(\cos\theta) + FQ_n^m(\cos\theta) \qquad (5.54)$$

where E and F are constants. It may be remarked that we need to restrict n to integer values to avoid a singularity at $\theta = 0$ and that we have to have $n > m$ because of the divergence problem of F at $\theta = 0$. Further, $\theta = \pi$ points to a singularity for $Q_n^m(\cos\theta)$. Hence for a finite solution for F we are led to the form

$$F(\theta) = EP_n^m(\cos\theta) \qquad (5.55)$$

where m is an integer. In such a case the series solution in (5.55) terminates at some desired order.

Hence a finite and single-valued general solution of $\phi(r, \theta, \phi)$ is

$$\phi(\vec{r}, t) = \sum_{n=0}^{\infty} \sum_{m=0}^{n-1} [A_n j_n(kr) + B_n y_n(kr)][a_m \cos(m\phi) + b_m \sin(m\phi)] P_n^m(\cos\theta) e^{-\frac{k^2}{\alpha^2} t}$$

$$(5.56)$$

Or equivalently

$$\phi(\vec{r}, t) = r^{-\frac{1}{2}} \sum_{n=0}^{\infty} \sum_{m=0}^{n-1} [A_n J_{n+\frac{1}{2}}(kr) + B_n J_{n+\frac{1}{2}}(kr)][a_m \cos(m\phi)$$

$$+ b_m \sin(m\phi)] P_n^m(\cos\theta) e^{-\frac{k^2}{\alpha^2} t} \quad (5.57)$$

where the boundary conditions determine the various unknown coefficients.

(c) Cylindrical polar coordinates (r, θ, z)

Let us proceed to seek solutions of the three-dimensional heat conduction equation in cylindrical polar coordinates (r, θ, z)

$$\phi_t(r, \theta, z, t) = \alpha^2 \nabla^2 \phi(r, \theta, z, t), \quad t > 0 \quad (5.58)$$

where from (3.45) ∇^2 is given by

$$\nabla^2 = \frac{\partial^2}{\partial r^2} + \frac{1}{r} \frac{\partial}{\partial r} + \frac{1}{r^2} \frac{\partial^2}{\partial \theta^2} + \frac{\partial^2}{\partial z^2} \quad (5.59)$$

and the ranges of the variables are $r \in [0, \infty)$, $\theta \in [0, 2\pi)$ and $z \in (\infty, \infty)$.

Let us assume that a solution exists in the variable-separated form

$$\phi(r, \theta, z, t) = R(r)F(\theta)G(z)T(t) \quad (5.60)$$

Substitution in (5.58) and using (5.59) gives the ODE

$$\frac{1}{G} \frac{d^2 G}{dz^2} + \frac{1}{R}(\frac{d^2 R}{dr^2} + \frac{1}{r} \frac{dR}{dr}) + \frac{1}{Fr^2} \frac{d^2 F}{d\theta^2} = \frac{1}{\alpha^2} \frac{1}{T} \frac{dT}{dt} \quad (5.61)$$

Since the left side is a function of z, r and θ and the right side is a function of t consistency requires that each side be equal to a constant which is set here as $-k^2$. Then we get the following solution for $T(t)$

$$T(t) = Ae^{-\alpha^2 k^2 t} \quad (5.62)$$

leaving for the remaining portion

$$\frac{1}{R}\left(\frac{d^2R}{dr^2} + \frac{1}{r}\frac{dR}{dr}\right) + \frac{1}{Fr^2}\frac{d^2F}{d\theta^2} + k^2 = -\frac{1}{G}\frac{d^2G}{dz^2} \tag{5.63}$$

In the above equation the left side is a function of r and θ and the right side is a function of z. Consistency requires that each side is equal to a constant which we set as $-q^2$. Then we get the following equation for $G(z)$

$$\frac{d^2G}{dz^2} - q^2z = 0 \tag{5.64}$$

whose general solution is

$$G(z) = Be^{qz} + Ce^{-qz} \tag{5.65}$$

where A and B are arbitrary constants. The solution corresponding to the first term in the right side blows up as $z \to \infty$ and so we need to set $B = 0$. This means that $G(z)$ reduces to

$$Z(z) = Ce^{-qz} \tag{5.66}$$

We are thus left with the differential equation

$$\frac{1}{R}\left(\frac{d^2R}{dr^2} + \frac{1}{r}\frac{dR}{dr}\right) + s^2 = -\frac{1}{Fr^2}\frac{d^2F}{d\theta^2} \tag{5.67}$$

where $s^2 = k^2 + q^2$.

In the above equation the left side is a function of r only and the right side is a function of θ only. So consistency requires that each side is equal to a constant which we set as w^2. Then we get the following pair of equations

$$\frac{d^2R}{dr^2} + \frac{1}{r}\frac{dR}{dr} + \left(s^2 - \frac{w^2}{r^2}\right)R = 0 \tag{5.68}$$

and

$$\frac{d^2F}{d\theta^2} + w^2F = 0 \tag{5.69}$$

Let us observe that with the substitution $\rho = sr$, the R-equation can be converted to

$$\frac{d^2R}{d\rho^2} + \frac{1}{\rho}\frac{dR}{d\rho} + \left(1 - \frac{w^2}{\rho^2}\right)R = 0 \tag{5.70}$$

where R is now a function of the new independent variable ρ. This equation can be readily recognized as the Bessel equation whose two linearly independent solutions are $J_w(\rho)$ and $Y_w(\rho)$ which are Bessel and Neumann functions respectively. Here w is to be looked upon as the order of the Bessel equation. Also note that $Y_w(\rho)$ diverges as $\rho \to 0$.

Reverting to the r variable, the solution of the R-equation is thus controlled by the form

$$\mathbb{R}(r) \propto J_w(sr) \tag{5.71}$$

because, from physics point of view, we are only interested in a finite solution as $r \to 0$.

Finally, the F-equation can be recognized as the harmonic oscillator equation whose solution is

$$F(\theta) = D\cos(w\theta) + S\sin(w\theta) \tag{5.72}$$

where D and S are arbitrary constants.

Hence the separable solution of $\phi(r, \theta, z, t)$ that remains finite as $r \to 0$ has the following form

$$\phi(r, \theta, z, t) = \Lambda J_w(sr)[M\cos(w\theta) + N\sin(w\theta)]e^{-(qz+\alpha^2 k^2 t)} \tag{5.73}$$

where Λ, M and N are constants of integration.

Example 5.5

A cylinder of radius a is defined between the planes $z = 0$ and $z = l$. Its plane faces are maintained at zero temperature while the temperature on the curved surface is $f(z)$. Find the steady symmetric temperature distribution within the cylinder.

Since we need to find the steady temperature distribution we can assume $\frac{\partial \phi}{\partial t} = 0$. Also because of the requirement of symmetric temperature distribution we can put $\frac{\partial \phi}{\partial \theta} = 0$. Hence the governing equation for the problem would read from (5.63)

$$\frac{1}{G}\frac{d^2 G}{dz^2} + \frac{1}{R}\left(\frac{d^2 R}{dr^2} + \frac{1}{r}\frac{dR}{dr}\right) = 0$$

Adopting a variable-separable form for ϕ

$$\phi(r, z) = R(r)G(z)$$

we are guided to the following ODE

$$\frac{1}{R}\left(\frac{d^2 R}{dr^2} + \frac{1}{r}\frac{dR}{dr}\right) = -\frac{1}{G}\frac{d^2 G}{dz^2} = k^2$$

where k^2 is a separation constant. This yields the following pair of equations

$$\frac{d^2 R}{dr^2} + \frac{1}{r}\frac{dR}{dr} - k^2 R = 0, \qquad \frac{d^2 G}{dz^2} + k^2 G = 0$$

Since we are interested in a finite solution, the solution of the R-equation is obviously governed by the Bessel function of order zero of the first kind $J_0(kr)$

while the boundary conditions $\phi(r, 0) = 0 = \phi(r, l)$ restrict the solution of the G-equation only to the sine part with $k = \frac{n\pi}{l}, n = 1, 2, \ldots$ Hence from the principle of superposition we can write the solution in the form

$$\phi(r, z) = \sum_{n=1}^{\infty} \Lambda_n J_0(\frac{n\pi r}{l}) \sin(\frac{n\pi z}{l})$$

where Λ_n is an overall constant.

To estimate Λ_n we make use of the given temperature function $f(z)$ at the surface. On putting $r = a$ we have

$$f(z) = \sum_{n=1}^{\infty} \Lambda_n J_0(\frac{n\pi a}{l}) \sin(\frac{n\pi z}{l})$$

We now multiply both sides by $\sin(\frac{n\pi z}{l})$, integrate between the limits 0 and l and make use of the orthogonality condition for the sine function. This gives for Λ_n

$$\Lambda_n = \frac{\frac{2}{l} \int_0^l dz f(z) \sin(\frac{n\pi z}{l})}{J_0(\frac{n\pi a}{l})}$$

and leads to the required solution

$$\phi(r, z) = \sum_{n=1}^{\infty} \frac{2}{l} [\int_0^l dz f(z) \sin(\frac{n\pi z}{l})] \frac{J_0(\frac{n\pi r}{l})}{J_0(\frac{n\pi a}{l})} \sin(\frac{n\pi z}{l})$$

5.5   Fundamental solution

We address the three-dimensional problem first. Let us express $\phi(\vec{r}, t)$ in terms of its Fourier transform $\tilde{\phi}(\vec{p}, t)$ and vice-versa through the pair of relations

$$\phi(\vec{r}, t) = (2\pi)^{-\frac{3}{2}} \int_{-\infty}^{\infty} \tilde{\phi}(\vec{p}, t) e^{i\vec{p}\cdot\vec{r}} d^3 p, \quad \tilde{\phi}(\vec{p}, t) = (2\pi)^{-\frac{3}{2}} \int_{-\infty}^{\infty} \phi(\vec{r}, t) e^{-i\vec{p}\cdot\vec{r}} d^3 r \tag{5.74}$$

where $\vec{p}$ is the transform variable. Substituting the first equation in (5.14) gives

$$(2\pi)^{-\frac{3}{2}} \int_{-\infty}^{\infty} d^3 p [\frac{\partial \tilde{\phi}(\vec{p}, t)}{\partial t} + \alpha^2 p^2 \tilde{\phi}(\vec{p}, t)] e^{i\vec{p}\cdot\vec{r}} = 0 \tag{5.75}$$

Multiplying now the left side of (5.75) by $e^{-i\vec{p}'\cdot\vec{r}}$ and performing integration over the entire space gives

$$(2\pi)^{-\frac{3}{2}} \int_{-\infty}^{\infty} d^3 p \int_{\infty}^{\infty} [d^3 r e^{i(\vec{p}-\vec{p}')\cdot\vec{r}}] [\frac{\partial \tilde{\phi}(\vec{p}, t)}{\partial t} + \alpha^2 p^2 \tilde{\phi}(\vec{p}, t)] = 0 \tag{5.76}$$

Using the Fourier transform of unity yielding the form of the delta function

$$\delta^3(\vec{p} - \vec{p'}) = (2\pi)^{-\frac{3}{2}} \int_{-\infty}^{\infty} d^3r e^{-i(\vec{p} - \vec{p'}) \cdot \vec{r}} \qquad (5.77)$$

(5.76) reduces to the form

$$\int_{-\infty}^{\infty} d^3p [\frac{\partial \tilde{\phi}(\vec{p}, t)}{\partial t} + \alpha^2 p^2 \tilde{\phi}(\vec{p}, t)] \delta^3 \vec{p} - \vec{p'}) = 0 \qquad (5.78)$$

(5.78) shows that $\tilde{\phi}(\vec{p}, t)$ obeys the equation

$$\frac{\partial \tilde{\phi}(\vec{p}, t)}{\partial t} + \alpha^2 p^2 \tilde{\phi}(\vec{p}, t) = 0 \qquad (5.79)$$

The solution of the PDE (5.79) can be expressed as

$$\tilde{\phi}(\vec{p}, t) = f(\vec{p}) e^{-\alpha^2 p^2 t} \qquad (5.80)$$

where f is an arbitrary function of $\vec{p}$. When (5.80) is substituted in the first equation of (5.74) we obtain for $\phi(\vec{r}, t)$

$$\phi(\vec{r}, t) = (2\pi)^{-\frac{3}{2}} \int_{-\infty}^{\infty} d^3p f(\vec{p}) e^{i\vec{p} \cdot \vec{r} - \alpha^2 p^2 t} \qquad (5.81)$$

We find an interesting result from here. At $t = 0$, setting $\phi(\vec{r}, 0) \equiv \xi(\vec{r})$ gives

$$\xi(\vec{r}) = (2\pi)^{-\frac{3}{2}} \int_{-\infty}^{\infty} d^3p f(\vec{p}) e^{i\vec{p} \cdot \vec{r}} \qquad (5.82)$$

whose inversion points to

$$f(\vec{p}) = (2\pi)^{-\frac{3}{2}} \int_{-\infty}^{\infty} d^3r \xi(\vec{r}) e^{-i\vec{p} \cdot \vec{r}} \qquad (5.83)$$

In other words, knowing the initial form $\xi(\vec{r})$ of $\phi(\vec{r}, t)$, we can determine $f(\vec{p})$ and hence from (5.81), know $\phi(\vec{r}, t)$ for all times.

Next, let us take the delta-function representation of $\xi(\vec{r})$ to write it as $\xi(\vec{r}) = \delta^3(\vec{r} - \vec{a})$ at some point $\vec{r} = \vec{a}$. Then by (5.77) $f(\vec{p})$ corresponds to

$$f(\vec{p}) = (2\pi)^{-\frac{3}{2}} e^{-i\vec{p} \cdot \vec{a}}, \quad \vec{a} \in \Re^3 \qquad (5.84)$$

It therefore follows from (5.81)

$$\phi(\vec{r}, t) = (2\pi)^{-\frac{3}{2}} \int_{-\infty}^{\infty} d^3p e^{i\vec{p} \cdot (\vec{r} - \vec{a}) - \alpha^2 p^2 t} \qquad (5.85)$$

To evaluate the integrals corresponding to the three components of $\vec{p}$, we use the following well-known results corresponding to the x, y and z components

$$\int_{\infty}^{\infty} dp_x e^{ip_x(x-a_x)-\alpha^2 p_x^2 t} = \sqrt{\frac{\pi}{\alpha^2 t}} e^{-\frac{(x-a_x)^2}{4\alpha^2 t}}, \quad t > 0 \qquad (5.86)$$

and

$$\int_{-\infty}^{\infty} dp_y e^{ip_y(y-a_y)-\alpha^2 p_y^2 t} = \sqrt{\frac{\pi}{\alpha^2 t}} e^{-\frac{(y-a_y)^2}{4\alpha^2 t}}, \quad t > 0,$$

$$\int_{-\infty}^{\infty} dp_z e^{ip_z(z-a_z)-\alpha^2 p_z^2 t} = \sqrt{\frac{\pi}{\alpha^2 t}} e^{-\frac{(z-a_z)^2}{4\alpha^2 t}}, \quad t > 0 \qquad (5.87)$$

These provide for $\phi(\vec{r}, t)$ the closed-form solution

$$\phi(\vec{r}, t) = (2\pi)^{-3} \left(\frac{\pi}{\alpha^2 t}\right)^{\frac{3}{2}} e^{-\frac{|\vec{r}-\vec{a}|^2}{4\alpha^2 t}}, \quad \vec{a} \in \Re^3, \quad t > 0 \qquad (5.88)$$

It is regarded as the fundamental or principal solution of the heat equation.

In one-dimension the solution reduces to

$$\phi(x, t) = \frac{1}{\sqrt{4\pi\alpha^2 t}} e^{-\frac{(x-a)^2}{4\alpha^2 t}}, \quad a \in \Re, \quad t > 0 \qquad (5.89)$$

We can also carry out a superposition to express $\phi(\vec{r}, t)$ in a general sense

$$\phi(x, t) = \int_{-\infty}^{+\infty} g(a) \frac{1}{\sqrt{4\pi\alpha^2 t}} e^{-\frac{(x-a)^2}{4\alpha^2 t}} da, \quad a \in \Re, \quad t > 0 \qquad (5.90)$$

where $g(a)$ is some continuous and bounded function of a. On carrying out differentiation under the sign of integration it is clear that $\phi(x, t)$ is a solution of the heat equation.

Some remarks are in order concerning the asymptotic behaviour of $\phi(x, t)$ with respect to t. For large values of both x and t, the ratios $\frac{x^2}{t}$ and $\frac{a^2}{t}$ behave respectively like ~ 1 and $\sim O(\epsilon)$, where $\epsilon << 1$. As a result the square root of the product behaves like $\frac{xa}{t} \sim O(\sqrt{\epsilon})$. It is thus straightforward to deduce from (5.90) that asymptotically

$$\phi(x, t) \to \frac{e^{-\frac{x^2}{4\alpha^2 t}}}{\sqrt{4\pi\alpha^2 t}} \int_{-\infty}^{+\infty} g(a) da, \quad a \in \Re \qquad (5.91)$$

implying

$$\phi(x, t) \to \frac{\tilde{g}(0)}{\sqrt{2t}} e^{-\frac{x^2}{4\alpha^2 t}}, \quad t > 0 \qquad (5.92)$$

where the appearance of $\tilde{g}(0)$ follows from the definition of the Fourier transform of $g(a)$ on setting the transform parameter to be zero. Thus asymptotically a damping nature of the solution is signaled with respect to the diffusion variable $\eta = \frac{x^2}{t}$.

5.6 Green's function

The Green's function is defined in the following manner. In three-dimensions it is given by

$$G^0(|\vec{r} - \vec{a}|, t) = (2\pi)^{-3}(\frac{\pi}{\alpha^2 t})^{\frac{3}{2}} e^{-\frac{|\vec{r}-\vec{a}|^2}{4\alpha^2 t}}, \quad a \in \Re, \quad t > 0 \tag{5.93}$$

while in one-dimension it has the form[1]

$$G^0(x - a, t) = \frac{1}{\sqrt{4\pi\alpha^2 t}} e^{-\frac{(x-a)^2}{4\alpha^2 t}}, \quad a \in \Re, \quad t > 0 \tag{5.94}$$

Note that the following properties for G^0 hold:

(i) $G^0(|\vec{r} - \vec{a}|, t)$ satisfies the homogeneous heat equation $\frac{\partial G^0}{\partial t} = \alpha^2 \nabla^2 G^0$. This is because for $t > 0$, G^0 coincides with $\phi(\vec{r}, t)$ according to (5.88) and since $\phi(\vec{r}, t)$ obeys the homogeneous heat equation, the same holds for G^0.

(ii) As $t \to 0^+$, the delta function limit is implied: $G^0(|\vec{r} - \vec{a}|, t) \to \delta^3(\vec{r} - \vec{a})$.

(iii) $G^0(|\vec{r} - \vec{a}|, t)$ obeys the integral

$$\int_{-\infty}^{+\infty} (\frac{1}{4\pi\alpha^2 t})^{\frac{3}{2}} e^{-\frac{|\vec{r}-\vec{a}|^2}{4\alpha^2 t}} \, dx dy dz = 1 \tag{5.95}$$

The proof easily follows by separating out the components and projecting the product form

$$e^{-\frac{|\vec{r}-\vec{a}|^2}{4\alpha^2 t}} = e^{-\frac{(x-a_x)^2}{4\alpha^2 t}} e^{-\frac{(y-a_y)^2}{4\alpha^2 t}} e^{-\frac{(z-a_z)^2}{4\alpha^2 t}} \tag{5.96}$$

The individual integrals with respect to x, y and z are evaluated by using the result

$$\int_{-\infty}^{+\infty} e^{-k(\zeta-a_\zeta)^2} \, d\zeta = \sqrt{\frac{\pi}{k}}, \quad \zeta = x, y, z \quad \text{and} \quad k = \frac{1}{4\alpha^2 t} \tag{5.97}$$

(iv) The link between $\phi(\vec{r}, t)$ and $\phi(\vec{r}, 0)$, where the latter is assumed to be known, is provided by the integral

$$\phi(\vec{r}, t) = \int \int \int dx' dy' dz' G^0(\vec{r} - \vec{r}', t) \phi(\vec{r}', 0), \quad t \geq 0 \tag{5.98}$$

---

[1] It corresponds to the Gaussian form discussed in Appendix A.

This is seen by operating with $(\frac{\partial}{\partial t} - \alpha^2 \nabla^2)$ on both sides of (5.98) and making use of the first property that G^0 satisfies the homogeneous wave equation. This in turn reflects that $\phi(\vec{r}, t)$ too is a solution of the homogeneous heat equation: $(\frac{\partial}{\partial t} - \alpha^2 \nabla^2) \phi(\vec{r}, t) = 0$.

(v) As $t \to 0$, $\phi(\vec{r}, t) \to \phi(\vec{r}, 0)$ on using property (ii) and (5.98) while as $r \to \infty$, $\phi(\vec{r}, t) \to 0$ since $G^0(\vec{r} - \vec{r}', t) \to 0$.

Consider the inhomogeneous case when $\phi(\vec{r}, t)$ solves the equation

$$\frac{\partial \phi}{\partial t} - \alpha^2 \nabla^2 \phi = \rho(\vec{r}, t) \tag{5.99}$$

where ρ is the inhomogeneous term. In such a case we can show that a particular solution of (5.99) enjoys an integral representation

$$\phi(\vec{r}, t) = \int \int \int dx' dy' dz' \int dt' \mathcal{G}^0(\vec{r} - \vec{r}', t - t') \rho(\vec{r}', t') \tag{5.100}$$

where $\mathcal{G}^0$ obeys the equation

$$(\frac{\partial}{\partial t} - \alpha^2 \nabla^2) \mathcal{G}^0(\vec{r} - \vec{a}, t - t_0) = \delta^3(\vec{r} - \vec{a}) \delta(t - t_0) \tag{5.101}$$

$\mathcal{G}^0$ is called the inhomgeneous Green's function.

To establish our claim, we define

$$\mathcal{G}^0(\vec{r} - \vec{a}, t - t_0) = G^0(\vec{r} - \vec{a}, t - t_0), \quad t > t_0 \tag{5.102}$$

$$\mathcal{G}^0(\vec{r} - \vec{a}, t - t_0) = 0, \quad t < t_0 \tag{5.103}$$

Then it is clear from the inhomogeneous equation (5.101) that for $t > t_0$ both the sides vanish, the left side due to G^0 being a solution of the homogeneous heat equation by property (i) and the right side due to the property of the delta function. Likewise for $t < t_0$, the left side vanishes by the definition of $\mathcal{G}^0$ and the right side due to the property of delta function.

Thus we have at hand a form of the particular solution of the inhomogeneous heat equation. Before we proceed to write down the general solution let us look at the behaviour of the particular solution as $t' \to 0^+$. If we integrate (5.101) over t' around an interval $(-\epsilon, +\epsilon)$, ϵ being an infinitesimal quantity, we find as ϵ proceeds to zero

$$\int_{-\epsilon}^{+\epsilon} dt' \frac{\partial}{\partial t'} \mathcal{G}^0(\vec{r} - \vec{a}, t') - \alpha^2 \int_{-\epsilon}^{+\epsilon} dt' \nabla^2 \mathcal{G}^0(\vec{r} - \vec{a}, t') = \int_{-\epsilon}^{+\epsilon} dt' \delta^3(\vec{r} - \vec{a}) \delta(t') = \delta^3(\vec{r} - \vec{a}) \tag{5.104}$$

where we have suppressed t_0. The first term in the left side when integrated out gives $G^0(\vec{r} - \vec{a}, +\epsilon)$ which, by property (ii) as mentioned earlier, acquires the form $\delta^3(\vec{r} - \vec{a})$ as ϵ goes to zero while the second term points to $-\alpha^2 \epsilon (\nabla^2 \mathcal{G}^0)|_{t \to 0+}$ which vanishes as ϵ goes to zero. Hence the behaviours of both sides are similar and consistency is maintained.

We are now in a position to write down the general solution for the inhomogeneous heat equation as the superposition of the general solution of the homogeneous part and the particular solution which we alluded to just now. In fact, if the initial condition on $\phi(\vec{r}, t)$ is given and $\rho(\vec{r}, t)$ is prescribed then from (5.98) and (5.100) the general solution for the inhomogeneous heat equation has the form

$$
\phi(\vec{r}, t) = \int_{-\infty}^{+\infty} \int_{-\infty}^{+\infty} \int_{-\infty}^{+\infty} dx' dy' dz' G^0(\vec{r} - \vec{r'}, t) \phi(\vec{r'}, 0)
$$
$$
+ \int_{-\infty}^{+\infty} \int_{-\infty}^{+\infty} \int_{-\infty}^{+\infty} dx' dy' dz' \int_0^{+\infty} dt' \mathcal{G}^0(\vec{r} - \vec{r'}, t - t') \rho(\vec{r'}, t'), \quad t > 0
$$
$$
(5.105)
$$

The proof is evident. The main points to be noted are that $\phi(\vec{r}, t)$ solves the inhomogeneous heat equation (5.99), that with the limit $t \to 0$, $\phi(\vec{r}, t)$ goes to $\phi(\vec{r}, 0)$ and further if both $\phi(\vec{r}, 0) = 0$ and $\rho(\vec{r}, t) = 0$ then $\phi(\vec{r}, t)$ also vanishes for all t. Moreover from the behaviour of $\mathcal{G}^0(\vec{r} - \vec{r'}, t)$ and $G^0(\vec{r} - \vec{r'}, t)$ for both the inequalities $t > 0$ and $t < 0$ already mentioned, $\phi(\vec{r}, t) \to 0$ as $r \to \infty$. All this also speaks for the required boundary conditions on $\phi(\vec{r}, t)$.

Since $\mathcal{G}^0(\vec{r} - \vec{r'}, t - t')$ vanishes for $t < t'$, we can also project $\phi(\vec{r}, t)$ given by the sum of the integrals

$$
\phi(\vec{r}, t) = \int_{-\infty}^{+\infty} \int_{-\infty}^{+\infty} \int_{-\infty}^{+\infty} dx' dy' dz' G^0(\vec{r} - \vec{r'}, t) \phi(\vec{r'}, 0)
$$
$$
+ \int_{-\infty}^{+\infty} \int_{-\infty}^{+\infty} \int_{-\infty}^{+\infty} dx' dy' dz' \int_0^t dt' \mathcal{G}^0(\vec{r} - \vec{r'}, t - t') \rho(\vec{r'}, t'), \quad t \geq 0
$$
$$
(5.106)
$$

as the general solution.

5.7 Summary

Starting with a brief generation of the reaction-diffusion equation which yields the form of the heat equation when the diffusion constant is treated as a constant, we examined the Cauchy problem for the heat equation focusing also on the uniqueness of the solution. We explored the maximum-minimum principle for the heat equation giving a detailed account of the proof by the approach of contradiction for both the inhomogeneous and the homogeneous forms. Subsequently we looked at the method of solution of the heat equation by separation of variables, We considered the individual case of the three-dimensional Cartesian coordinates which also guided us to the simpler versions of the one-dimensional and two-dimensional cases. We also derived the separable solutions of the three-dimensional polar coordinates and cylindrical polar coordinates. Seeking the fundamental solution of the heat equation was our next topic of inquiry when we extracted the form of the standard representation formula. We then solved the problem for the Green's function that allows us to write the general solution for the inhomogeneous heat equation as the superposition of the general solution of the homogeneous equation along with a particular solution of the inhomogeneous equation both in their typical integral representations.

Exercises

1. Consider the one-dimensional parabolic equation

$$\phi_t(x,t) = \phi_{xx}(x,t), \quad -\infty < x < \infty, \quad t > 0$$

obeying the condition

$$\phi(x,0) = f(x), \quad -\infty < x < \infty$$

Show that the following integral

$$\phi(x,t) = \frac{1}{2\sqrt{\pi t}} \int_{-\infty}^{+\infty} \phi(w) e^{-\frac{(x-w)^2}{4w}} dw, \quad t > 0$$

is a solution, where $\phi(y)$ may be assumed to be bounded in $-\infty < y < \infty$.

2. Solve the one-dimensional equation of heat conduction as given by

$$\phi_t(x,t) = \phi_{xx}(x,t), \quad 0 < x < \infty$$

where x denotes the distance measured along a semi-infinite rod from a fixed point, $\phi(x,t)$ is the temperature at any point of the rod at $t > 0$. Assume that the equation is subject to the initial condition

$$\phi(x,0) = q(x)$$

along with the boundary condition

$$\phi(0,t) = 0$$

Assume also that ϕ and all of its x- derivatives vanish asymptotically.
    Examine the specific case of heat flow in a bar of length l when $q(x)$ is defined by

$$q(x) = \begin{cases} 1 & x \in [1,2] \\ 0 & \text{otherwise} \end{cases} \tag{5.107}$$

and show that the profile is

$$\phi(x,t) = \int_0^\infty \frac{2}{\pi p} (\cos p - \cos 2p) e^{-p^2 t} \sin px \, dp$$

3. Solve the following initial-boundary value problem of heat flow in a bar of length l

$$\phi_t(x,t) - \alpha^2 \phi_{xx}(x,t) = q(x), \quad 0 < x < l, \quad t > 0$$

whose end points are maintained at zero temperature

$$\phi_x(0,t) = 0, \quad \phi_x(l,t) = 0, \quad \phi(x,0) = s(x), \quad 0 < x < l$$

Show that the solution is

$$\phi_x(x,t) = \int_0^l \left(\frac{2}{l} \sum_{n=1}^{\infty} e^{\frac{-n^2 \pi^2 t}{l^2}} \sin\frac{n\pi x}{l} \sin\frac{n\pi\xi}{l}\right) s(\xi) d\xi$$

4. Solve the following initial-boundary value problem of heat flow in a bar of length l

$$\phi_t(x,t) - \alpha^2 \phi_{xx}(x,t) = q(x), \quad 0 < x < l, \quad t > 0$$

whose end points are maintained at constant temperature

$$\phi_x(0,t) = \gamma, \quad \phi_x(l,t) = \delta, \quad \phi(x,0) = s(x)$$

where γ and δ are constants.

5. Solve the one-dimensional equation of heat conduction

$$\phi_t(x,t) = \phi_{xx}(x,t), \quad 0 < x < l$$

whose end-points are maintained at zero-temperature

$$\phi(0,t) = 0, \quad \phi(1,t) = 0 \quad \text{for all } t$$

along with the initial conditions

$$\phi(x,0) = x, \quad 0 \le x \le \frac{l}{2} \quad \text{and} \quad \phi(x,0) = l - x, \quad \frac{l}{2} \le x \le l$$

6. Suppose that the one-dimensional parabolic equation

$$\phi_t(x,t) = \alpha^2 \phi_{xx}(x,t), \quad t > 0$$

has a positive solution $\phi(x,t)$. Show that the function ψ given by the form $-2\alpha^2 \frac{\phi_x}{\phi}$ solves Burger's equation[2]

---

[2]The general form of Burger's equation is $\psi_t(x,t) + \psi\psi_x = D\psi_{xx}(x,t)$. The term in the right side is the diffusion term.

$$\psi_t + \psi\psi_x = \alpha^2 \psi_{xx}, \quad t > 0$$

7. Consider the two-dimensional parabolic equation

$$\phi_t(x, y, t) = \phi_{xx}(x, y, t) + \phi_{yy}(x, y, t), \quad t > 1$$

where it is given that $\phi(x, y, 1) = 1 - (x^2 + y^2)^2$. Show that

$$\phi(x, y, t) = 1 - (x^2 + y^2)^2 - 16(x^2 + y^2)(t - 1) - 32(t - 1)^2$$

solves the equation.

8. The two-dimensional diffusion equation is given by

$$\phi_t = D(\phi_{xx} + \phi_{yy}), \quad D > 0$$

Applying the transformation to polar coordinates $x = r\cos\theta, y = r\sin\theta$, obtain the form

$$\phi_t = D\left(\phi_{rr} + \frac{1}{r}\phi_r\right)$$

Show that the functions $e^{-t}J_k(r)\cos(k\phi)$ and $e^{-t}J_k(r)\sin(k\phi)$, $k = 0, 1, 2, ...$ satisfy the above equation for $D = 1$. Here J_k is Bessel function of integral order k.

9. Consider a homogeneous solid sphere of radius a possessing an initial temperature distribution $f(r)$, r being the distance from the centre of the sphere. Solve for the temperature distribution $T(r)$ in the sphere from the equation

$$\phi_t = D\left(T_{rr} + \frac{2}{r}T_r\right)$$

where it may be assumed that the surface of the sphere is maintained at zero temperature.

10. Determine the temperature $T(r, t)$ for an infinite cylinder $0 \leq r \leq a$ given that the initial temperature is $T(r, 0) = f(r)$ and that the surface $r = a$ is maintained at zero degree.

Chapter 6

Solving PDEs by integral transform method

The integral transform method is an effective tool of reducing the complexity of PDEs to a clear tractable form that allows many problems of mathematical physics to be tackled straightforwardly. In this chapter we will be primarily interested in solving problems of PDEs that requires an employment of either the Fourier transform or the Laplace transform technique. The essence of the method of either transform is to reduce a given PDE into an ODE in a somewhat similar spirit as that of the separation of variables. However, here the main difference lies in the fact that while in the case of the Fourier transform, the domain of the differential operator is defined over the full-line $(-\infty, +\infty)$ and hence the method is applicable when the given function is defined over an entire real line equipped with appropriate boundary conditions, for the Laplace transform, since the domain of the differential operator is only the half-line $(0, \infty)$, the method becomes relevant when we are dealing with an initial value problem. Such a half-line-full-line contrast of the two transforms is due to the fact that it often happens that Fourier transforms of certain classes of functions may not exist, such as for example x^2, and a control factor is necessary to be appended to the function to ensure better convergence behaviour on at least on one side of the interval. We have discussed and collected some useful analytical results of the Fourier and Laplace transforms in Appendix B and Appendix C respectively. In the following we begin first with the Fourier transform method.

6.1   Solving by Fourier transform method

Let us apply the Fourier transform method to deduce solutions of some typical PDEs that we generally encounter in standard physical situations. The following illustrative examples will make the strategy clear.

Example 6.1

Solve the one-dimensional heat conduction equation

$$\phi_t(x,t) = \alpha^2 \phi_{xx}(x,t), \quad -\infty < x < \infty, \quad t > 0 \tag{6.1}$$

where $\phi(x,t)$ is the temperature at any point of the rod and x is the distance measured along the rod from a fixed point, subject to the initial condition

$$\phi(x,0) = \phi_0(x) \tag{6.2}$$

To solve by the Fourier transform method we assume that the relevant Fourier transform exists and the Dirichlet's conditions hold for its convergence (see Appendix B). Multiplying both sides of the given heat equation by $(2\pi)^{-\frac{1}{2}} e^{-ipx}$ and integrating over the full range $(-\infty, +\infty)$ we have

$$
\begin{aligned}
(2\pi)^{-\frac{1}{2}} \int_{-\infty}^{+\infty} \phi_t(x,t) e^{-ipx} dx &= (2\pi)^{-\frac{1}{2}} \alpha^2 \int_{-\infty}^{+\infty} \phi_{xx}(x,t) e^{-ipx} dx \\
&= -(2\pi)^{-\frac{1}{2}} \alpha^2 \int_{-\infty}^{+\infty} \phi_x(x,t)[(ip)]e^{-ipx}]dx \\
&= (2\pi)^{-\frac{1}{2}} \alpha^2 (ip)^2 \int_{-\infty}^{+\infty} \phi(x,t) e^{-ipx} dx \\
&= -\alpha^2 p^2 \tilde{\phi} \tag{6.3}
\end{aligned}
$$

where $\tilde{\phi}$ is the Fourier transform of ϕ and in the second and third steps we integrated by parts and assumed ϕ and $\phi_x \to 0$ as $x \to \pm\infty$. Taking the time-derivative out of the integral, the left side gets converted to an ordinary derivative and we get the form of an ODE

$$\frac{d\tilde{\phi}}{dt} = -\alpha^2 p^2 \tilde{\phi} \tag{6.4}$$

Next, we take the Fourier transform of (6.2) with respect to x and obtain

$$\tilde{\phi}(p,0) = (2\pi)^{-\frac{1}{2}} \int_{-\infty}^{+\infty} \phi_0(x) e^{-ipx} dx \equiv \tilde{\phi}_0 \tag{6.5}$$

As a result we can express the solution of (6.4) as

$$\tilde{\phi}(p,t) = \tilde{\phi}_0 e^{-\alpha^2 p^2 t} \tag{6.6}$$

where we have accounted for (6.5). Notice that the nature of $\tilde{\phi}(p,t)$ is of decaying type. The task now is to invert (6.4) to the coordinate representation of $\tilde{\phi}$.

To proceed from here we make use of the fact that the quantity $e^{-\alpha^2 p^2 t}$ can be expressed as a Fourier transform

$$e^{-\alpha^2 p^2 t} = F[\frac{1}{a\sqrt{2t}}e^{-\frac{x^2}{4t\alpha^2}}]$$ (6.7)

This means that the right side of (6.6) is basically a product of two Fourier transforms i.e.

$$\tilde{\phi}(p,t) = \tilde{\phi}_0 F[\frac{1}{a\sqrt{2t}}e^{-\frac{x^2}{4t\alpha^2}}]$$ (6.8)

By the convolution theorem of Fourier transform $\phi(x,t)$ can be provided an integral representation

$$\phi(x,t) = \frac{1}{2\alpha\sqrt{\pi t}}\int_{\infty}^{\infty} d\xi \phi_0(\xi)e^{-\frac{(x-\xi)^2}{4t\alpha^2}}$$ (6.9)

and gives the temperature distribution at any point x in the rod. It can be solved completely if we know the form of ϕ_0.

Example 6.2

Find the solution of the two-dimensional Laplace's equation in the half-space

$$\phi_{xx}(x,y) + \phi_{yy}(x,y) = 0, \quad x \in (-\infty,\infty), \quad y \geq 0$$ (6.10)

which is governed by the boundary condition

$$\phi(x,0) = f(x), \quad x \in (-\infty,\infty),$$ (6.11)

along with $\phi(x,y) \to 0$ for both $|x|$ and $y \to \infty$.

Multiplying both sides of the Laplace's equation by $\frac{1}{\sqrt{2\pi}}e^{-ipx}$ and integrating over $(-\infty,\infty)$ with respect to the variable x gives the conversion to an ODE

$$\frac{d^2\tilde{\phi}}{dy^2} - p^2\tilde{\phi} = 0$$ (6.12)

where we have assumed ϕ and $\phi_x \to 0$ as $x \to \pm\infty$, $\tilde{\phi}$ stands for the Fourier transform

$$\tilde{\phi}(p,y) = \frac{1}{\sqrt{2\pi}}\int_{-\infty}^{\infty} \phi(x,y)e^{-ipx}dx$$ (6.13)

and p is the transform parameter.

Taking Fourier transform of the boundary condition (6.11) gives

$$\tilde{\phi}(p,0) = \tilde{f}(p)$$ (6.14)

where $\tilde{f}(p)$ is the Fourier transform of $f(x)$

$$\tilde{f}(p) = \frac{1}{\sqrt{2\pi}} \int_{-\infty}^{\infty} f(x)e^{-ipx}\,dx \tag{6.15}$$

Further

$$\tilde{\phi}(p, y) \quad \to 0 \quad \text{as} \quad y \to \infty \tag{6.16}$$

Now the general solution of the differential equation (6.12) is of the form

$$\hat{\phi}(p, y) = Ae^{-py} + Be^{py} \tag{6.17}$$

where A and B are are arbitrary constants. Because of (6.16), we have to impose $B = 0$ for $p > 0$ and $A = 0$ for $p < 0$. Combining we can write

$$\hat{\phi}(p, y) = \hat{f}(p)e^{-|p|y} \quad \text{for all p} \tag{6.18}$$

Setting $\hat{g}(p, y) = e^{-|p|y}$, the inversion formula of Fourier transform gives for $g(x, y)$

$$g(x, y) = \frac{1}{\sqrt{2\pi}} \int_{-\infty}^{\infty} e^{-|p|y}e^{ipx}\,dp \tag{6.19}$$

Decomposing the integral range into two parts $(-\infty, 0)$ and $(0, \infty)$, $g(x, y)$ turns out to be

$$g(x, y) = \frac{1}{\sqrt{2\pi}} \int_{0}^{\infty} e^{-p(y+ix)}\,dp + \frac{1}{\sqrt{2\pi}} \int_{0}^{\infty} e^{-p(y-ix)}\,dp = \frac{2y}{\sqrt{2\pi}(x^2 + y^2)} \tag{6.20}$$

where the integrals are straightforward to evaluate.

Since by (6.18)

$$\hat{\phi}(p, y) = \hat{f}(p)\hat{g}(p, y) \tag{6.21}$$

the convolution theorem of Fourier transform gives (see Appendix B)

$$\phi(x, y) = \frac{1}{\sqrt{2\pi}} \int_{-\infty}^{\infty} f(\xi)g(x - \xi, y)\,d\xi \tag{6.22}$$

which, by using (6.20), points to the integral

$$\phi(x, y) = \frac{y}{\pi} \int_{-\infty}^{\infty} \frac{f(\xi)d\xi}{y^2 + (x - \xi)^2} \tag{6.23}$$

The exact evaluation of (6.14) depends on the specific form of $f(x)$.

Example 6.3: D'Alembert's solution

Solve by employing the method of Fourier transform the following Cauchy problem of free vibration of an infinitely stretched string

$$\phi_{tt}(x,t) = c^2 \phi_{xx}(x,t), \quad -\infty \le x < \infty, \quad t \ge 0 \tag{6.24}$$

subject to the initial conditions

$$\phi(x,0) = f(x) \quad \text{and} \quad \phi_t(x,0) = g(x) \tag{6.25}$$

Multiplying both sides of (6.24) by $\frac{1}{\sqrt{2\pi}} e^{-ipx}$, integrating with respect to x over $(-\infty, \infty)$ and assuming the vanishing limits $\phi, \phi_x \to 0$ as $x \to \pm\infty$ gives

$$\frac{d^2\tilde{\phi}}{dt^2} = -c^2 p^2 \tilde{\phi} \tag{6.26}$$

where $\tilde{\phi}(t,p)$ is the Fourier transform of ϕ with respect to x

$$\tilde{\phi}(t,p) = \frac{1}{\sqrt{2\pi}} \int_{-\infty}^{\infty} \phi(x,t) e^{-ipx} dx \tag{6.27}$$

The general solution of the above equation is given by

$$\tilde{\phi}(t,p) = A e^{ipct} + B e^{-ipct} \tag{6.28}$$

where A and B are arbitrary constants and need to be determined from the given conditions (6.25).

Defining the Fourier transforms of f and g to be

$$\tilde{f}(p) = \frac{1}{\sqrt{2\pi}} \int_{-\infty}^{\infty} f(x) e^{-ipx} dx, \quad \tilde{g}(p) = \frac{1}{\sqrt{2\pi}} \int_{-\infty}^{\infty} g(x) e^{-ipx} dx \tag{6.29}$$

along with their corresponding inversions

$$f(x) = \frac{1}{\sqrt{2\pi}} \int_{-\infty}^{\infty} \tilde{f}(p) e^{ipx} dp, \quad g(x) = \frac{1}{\sqrt{2\pi}} \int_{-\infty}^{\infty} \tilde{g}(p) e^{ipx} dp \tag{6.30}$$

we obtain from the initial conditions

$$\tilde{\phi} = \tilde{f}, \quad \frac{d\tilde{\phi}}{dt} = \tilde{g} \quad \text{at} \quad t = 0 \tag{6.31}$$

From (6.26) and (6.31) the following estimates of A and B then follow

$$A = \frac{1}{2}\left(\tilde{f} - i\frac{\tilde{g}}{pc}\right), \quad B = \frac{1}{2}\left(\tilde{f} + i\frac{\tilde{g}}{pc}\right) \tag{6.32}$$

Substituting them in (6.28) we are guided to the form

$$\phi(x,t) = \frac{1}{2}\frac{1}{\sqrt{2\pi}}\int_{-\infty}^{\infty}\tilde{f}(p)[e^{-ip(x-ct)} + e^{-ip(x+ct)}]dp$$

$$- \frac{i}{2}\frac{1}{\sqrt{2\pi}}\int_{-\infty}^{\infty}\frac{\tilde{g}(p)}{p}[e^{-ip(x-ct)} - e^{-ip(x+ct)}]dp \qquad (6.33)$$

Since using (6.30) it is possible to express for $f(x+ct)$ and $f(x-ct)$ the respective integrals

$$f(x+ct) = \frac{1}{2}\frac{1}{\sqrt{2\pi}}\int_{-\infty}^{\infty}\tilde{f}(p)[e^{-ip(x+ct)}dp \qquad (6.34)$$

$$f(x-ct) = \frac{1}{2}\frac{1}{\sqrt{2\pi}}\int_{-\infty}^{\infty}\tilde{f}(p)[e^{-ip(x-ct)}dp \qquad (6.35)$$

and for the integral $\int_{x-ct}^{x+ct} g(u)du$ the expression

$$\int_{x-ct}^{x+ct} g(u)du = -\frac{i}{\sqrt{2\pi}}\int_{-\infty}^{\infty}\frac{\tilde{g}(p)}{p}[e^{-ip(x-ct)} - e^{-ip(x+ct)}]dp \qquad (6.36)$$

the final form of $\phi(x,t)$ is given by

$$\phi(x,t) = \frac{1}{2}[f(x+ct) + f(x-ct)] + \frac{1}{2c}\int_{x-ct}^{x+ct} g(u)du \qquad (6.37)$$

which coincides with the D'Alembert's solution already determined in Chapter 4.

Example 6.4

Solve the linear flow of heat problem in a semi-infinite rod

$$\phi_t(x,t) = \alpha^2\phi_{xx}(x,t), \quad -\infty < x < \infty \qquad (6.38)$$

where x denotes the distance measured along the rod from a fixed point, $\phi(x,t)$ is the temperature at any point of the rod at $t > 0$. It is given that the equation is subject to the initial condition

$$\phi(x,0) = q(x), \quad x > 0 \qquad (6.39)$$

and the following boundary condition at $x = 0$

$$\phi(0,t) = r(t), \quad t > 0 \qquad (6.40)$$

Further $\phi- > 0$ as $x- > \infty$ for $t > 0$. Find the solution when $r(t) = \phi_o$, where ϕ_0 is a constant.

Here taking Fourier sine transform would be worthwhile

$$\phi_s(p,t) = \sqrt{\frac{2}{\pi}}\int_0^{\infty}\phi(x,t)\sin pxdx \qquad (6.41)$$

Multiplying both sides of (6.24) by $\sqrt{\frac{2}{\pi}}\sin px$ and integrating from 0 to ∞, the equation is converted to the form

$$\frac{\partial\phi_s(p,t)}{\partial t} = \alpha^2[\sqrt{\frac{2}{\pi}}pr(t) - p^2\phi_s] \tag{6.42}$$

where we have used the formula $(B.24)$ given in Appendix B

$$[f_{xx}]_s = \sqrt{\frac{2}{\pi}}pf(0) - p^2 f_s \tag{6.43}$$

the suffix s indicating the sine transform and $f(x,t)$ is any given function.

If we now apply the initial condition in (6.39), we obtain from (6.41)

$$\phi_s(p,0) = \sqrt{\frac{2}{\pi}} \int_0^\infty q(x)\sin px dx \tag{6.44}$$

Solving (6.42) in the presence of (6.44) yields $\phi_s(p,t)$. Hence $\phi(x,t)$, on taking the inverse sine transform, is given by

$$\phi(x,t) = \int_0^\infty \phi_s(p,t)\sin px dp \tag{6.45}$$

where $\phi_s(p,t)$ is known from the following argument. When $r(t) = \phi_0$, (6.28) becomes

$$\frac{\partial\phi_s(p,t)}{\partial t} + \alpha^2 p^2\phi_s(p,t) = \alpha^2 p\sqrt{\frac{2}{\pi}}\phi_0 \tag{6.46}$$

which has the solution

$$\phi_s(p,t) = \sqrt{\frac{2}{\pi}}\frac{\phi_0}{p}(1 - e^{-\alpha^2 p^2 t}) \tag{6.47}$$

Employing (6.47), $\phi(x,t)$, from (6.45), turns out to be

$$\phi(x,t) = \sqrt{\frac{2}{\pi}}\phi_0 \int_0^\infty \frac{\sin px}{p}(1 - e^{-\alpha^2 p^2 t})dp \tag{6.48}$$

which is the integral representation of the solution. We can express $\phi(x,t)$ in terms of the error function by observing that the latter is defined by

$$erf(y) = \frac{2}{\sqrt{\pi}} \int_0^y e^{-u^2} du \tag{6.49}$$

and obeys the following property

$$\int_0^\infty e^{-u^2}\frac{\sin(2uy)}{u}du = \frac{\pi}{2}erf(y) \tag{6.50}$$

A little manipulation then shows that $\phi(x,t)$ can be expressed in the form

$$\phi(x,t) = \sqrt{\frac{2}{\pi}}\phi_0[\frac{\pi}{2} - \frac{\pi}{2}erf(\frac{x}{2\alpha\sqrt{t}})] \tag{6.51}$$

which can also be recast as

$$\phi(x,t) = \sqrt{\frac{\pi}{2}}\phi_0(1 - \frac{2}{\sqrt{\pi}}\int_0^{\frac{x}{2\alpha\sqrt{t}}} e^{-u^2} du) \tag{6.52}$$

6.2   Solving by Laplace transform method

We now proceed to find the solution of a class of initial-value problems by employing the Laplace transform method. We focus here on the one-dimensional problems only. In such models the governing PDE invariably contains a function having the time parameter t as one of the independent variables with a prescribed initial condition. By looking for the Laplace transform over the interval $(0,\infty)$, we can convert the given equation along with the initial-boundary conditions into an ODE of a complex-valued function in terms of the Laplace variable s. Once this equation is solved we can make use of the inversion formula outlined in Appendix C to arrive at the desired solution in a certain tractable form. We note in passing that for the higher-dimensional PDEs one may have to take recourse to the application of the Laplace transform more than once and subsequently go for an adequate number of inversions. It sometimes could happen that the Laplace transform method may not prove to be suitable for a given PDE in extracting its solution in a closed form. In such cases the limiting behaviour of the solution, subject to certain conditions, may be extracted. A couple of illustrative examples have been considered in this regard as well.

Example 6.5

Solve the heat conduction equation by employing the method of Laplace transform

$$\phi_t(x,t) = \alpha^2\phi_{xx}(x,t), \quad x \geq 0, \quad t > 0 \tag{6.53}$$

subject to the initial-boundary conditions

$$\phi(x,0) = 0, \quad x > 0, \quad \phi(0,t) = f(t) \tag{6.54}$$

Multiplying both sides of the diffusion equation by e^{-st} and integrating with respect to the variable t over $(0, \infty)$ we get

$$s \int_0^\infty e^{-st} \phi \, dt = \alpha^2 \frac{d^2}{dx^2} \int_0^\infty e^{-st} \phi \, dt \tag{6.55}$$

where the boundary term in the left side has been dropped by employing the vanishing initial condition on ϕ and also noting that the latter does not show any singular behaviour at $t \to \infty$. We thus arrive at the ODE

$$\bar{\phi}_{xx} = \frac{s}{\alpha^2} \bar{\phi} \tag{6.56}$$

where $\bar{\phi}$ is the Lapalce transform of ϕ

$$\bar{\phi}(x, s) = \int_0^\infty e^{-st} \phi(x, t) dt \tag{6.57}$$

with respect to t.

The general solution of the above ODE is of the type

$$\bar{\phi}(x, s) = A e^{-\frac{\sqrt{s}x}{\alpha}} + B e^{\frac{\sqrt{s}x}{\alpha}} \tag{6.58}$$

in which we have to set $B = 0$ to avoid blowing up of the solution at $x \to \infty$. We are thus led to the reduced form

$$\bar{\phi}(x, s) = A e^{-\frac{\sqrt{s}x}{\alpha}} \tag{6.59}$$

in terms of an overall constant A.

Taking the Laplace transform of the second boundary condition in (6.54) gives

$$\bar{\phi}(0, s) = F(s) \tag{6.60}$$

where $F(s)$ is the Laplace transform of $f(t)$. Comparing (6.59) and (6.60) fixes A to be $A = F(s)$. We thus have for $\bar{\phi}$ the form

$$\bar{\phi}(x, s) = F(s) e^{-\frac{\sqrt{s}x}{\alpha}} \tag{6.61}$$

We notice that the right side is a product of two Laplace transforms of which the second one stands for

$$e^{-\frac{x\sqrt{s}}{\alpha}} = L\left[\frac{x}{2\alpha t \sqrt{\pi t}} e^{-\frac{x^2}{4\alpha^2 t}}\right] \tag{6.62}$$

where we have utilized the following standard result (see the table in Appendix C)

$$e^{-a\sqrt{s}} = L\left[\frac{a}{2t\sqrt{\pi t}} e^{-\frac{a^2}{4t}}\right] \tag{6.63}$$

by scaling the parameter a as $a \to \frac{x}{\alpha}$.

The following integral form for $\phi(x,t)$ is thus suggested from (6.61) and (6.62) by employing the convolution integral (C.65) of Laplace theorem (see Appendix C)

$$\phi(x,t) = \int_0^t f(\tau) \frac{x}{2\alpha\sqrt{\pi}(t-\tau)\sqrt{t-\tau}} e^{-\frac{x^2}{4\alpha^2(t-\tau)}} d\tau \tag{6.64}$$

It also corresponds to

$$\phi(x,t) = \frac{x}{2\alpha\sqrt{\pi}} \int_0^t \frac{f(\tau)}{(t-\tau)\sqrt{t-\tau}} e^{-\frac{x^2}{4\alpha^2(t-\tau)}} d\tau \tag{6.65}$$

by taking the variable x out of the integral sign. (6.64) or (6.65) provides the required solution for $\phi(x,t)$.

Example 6.6

Solve the heat equation in the closed interval $[0,a]$, $a > 0$ by employing the method of Laplace transform

$$\phi_t(x,t) = \alpha^2 \phi_{xx}(x,t), \quad 0 \le x \le a, \quad t > 0 \tag{6.66}$$

subject to the initial condition

$$\phi(x,0) = 0, \quad 0 \le x \le a \tag{6.67}$$

and the boundary conditions

$$\phi(0,t) = 0, \quad \phi(a,t) = \phi_0, \quad t > 0 \tag{6.68}$$

As in the previous problem we multiply both sides of the given PDE by e^{-st}, integrate with respect to the variable t over $(0,\infty)$ and drop the boundary term on applying the initial condition (6.67). Thus we arrive at the ODE

$$\bar{\phi}_{xx} = \frac{s}{\alpha^2} \bar{\phi} \tag{6.69}$$

where $\bar{\phi}$ is the Laplace transform of ϕ with respect to the variable t

$$\bar{\phi}(x,s) = \int_0^\infty e^{-st} \phi(x,t) dt \tag{6.70}$$

The general solution of (6.69) is

$$\bar{\phi}(x,s) = A e^{\frac{\sqrt{s}x}{\alpha}} + B e^{-\frac{\sqrt{s}x}{\alpha}} \tag{6.71}$$

where A and B are the arbitrary constants.

Next, taking the Laplace transform of the two boundary conditions in (6.68) we have

$$\bar{\phi}(0, s) = 0, \quad \bar{\phi}(a, s) = \frac{\phi_0}{s} \tag{6.72}$$

By bringing to terms with the above conditions we can solve for A and B to obtain

$$A = \frac{\phi_0}{2s \sinh(\frac{\sqrt{s}a}{\alpha})}, \quad B = -\frac{\phi_0}{2s \sinh(\frac{\sqrt{s}a}{\alpha})} \tag{6.73}$$

Hence $\bar{\phi}(x, s)$ assumes the form

$$\bar{\phi}(x, s) = \frac{\phi_0}{s} \frac{\sinh(\frac{\sqrt{s}x}{\alpha})}{\sinh(\frac{\sqrt{s}a}{\alpha})} \tag{6.74}$$

Now, through inversion, we can write

$$\phi(x, t) = \frac{1}{2i\pi} \int_{r-i\infty}^{r+i\infty} e^{st} \frac{\phi_0}{s} \frac{\sinh(\frac{\sqrt{s}x}{\alpha})}{\sinh(\frac{\sqrt{s}a}{\alpha})} \tag{6.75}$$

The singularities in the integrand of (6.75) are seen to be located at

$$s = 0, \quad \frac{ia\sqrt{s}}{\alpha} \quad \Rightarrow \quad s = -\frac{\alpha^2 n^2 \pi^2}{a^2}, n = 0, 1, 2, ... \tag{6.76}$$

The results for the residues are as follows

$$\text{at} \quad s = 0: \quad \frac{x\phi_0}{a} \tag{6.77}$$

and

$$\text{at} \quad s = -\frac{\alpha^2 n^2 \pi^2}{a^2}, n = 1, 2, ... : \quad \frac{2\phi_0}{n\pi}(-1)^n e^{-\frac{\alpha^2 n^2 \pi^2 t}{a^2}} \sin \frac{n\pi x}{a} \tag{6.78}$$

Therefore

$$\phi(x, t) = \phi_0 [\frac{x}{a} + \frac{2}{\pi} \sum_{n=1}^{\infty} \frac{(-1)^n}{n} e^{-\frac{\alpha^2 n^2 \pi^2 t}{a^2}} \sin \frac{n\pi x}{a}] \tag{6.79}$$

solves the given PDE satisfying the prescribed conditions of the problem.

Example 6.7

Find the asymptotic expansion of $\phi(t)$ when $\phi(t)$ satisfies the heat equation

$$\phi_t(x,t) = \alpha^2 \phi_{xx}(x,t), \quad x \geq 0, \quad t > 0 \tag{6.80}$$

subject to the initial condition

$$\phi(x,0) = 0, \quad x \geq 0 \tag{6.81}$$

and the boundary condition

$$\phi_x(0,t) = h(\phi - \phi_0), \quad t > 0 \tag{6.82}$$

where h and ϕ_0 are constants.

The Laplace transform of the given equation with respect to t yields, as usual, the ODE

$$\bar{\phi}_{xx} = \frac{s}{\alpha^2} \bar{\phi} \tag{6.83}$$

in which the given initial condition is used and $\bar{\phi}(x,s)$ denotes the Laplace transform integral of $\phi(x,t)$ with respect to the variable t

$$\bar{\phi}(x,s) = \int_0^\infty e^{-st} \phi(x,t) dt \tag{6.84}$$

The general solution of (6.83) is of course similar to (6.71) as provided in terms of the two arbitrary constants A and B. Requiring a finite solution of ϕ at $x \to \infty$ restricts $B = 0$ and we have

$$\bar{\phi}(x,s) = A e^{-\frac{\sqrt{s}x}{\alpha}} \tag{6.85}$$

in terms of A only.

Taking next the Laplace transform of the boundary condition (6.82) we have at $x = 0$

$$\bar{\phi}_x(0,s) = h(\bar{\phi}|_{x=0} - \frac{\phi_0}{s}) = h(A - \frac{\phi_0}{s}) \tag{6.86}$$

Seeking consistency between (6.85) and (6.86) at $x = 0$ boundary fixes A

$$A = \frac{h\phi_0}{s(h + \sqrt{\frac{s}{k}})} \tag{6.87}$$

Thus $\bar{\phi}(x,s)$ acquires the form

$$\bar{\phi}(x,s) = \frac{h\phi_0}{s(h + \sqrt{\frac{s}{k}})} e^{-\frac{\sqrt{s}x}{\alpha}} \tag{6.88}$$

Note that the branch point $s = 0$ is the only singularity of $\bar{\phi}(x, s)$ since for $h > 0$ the quantity $h + \sqrt{\frac{s}{k}}$ cannot vanish. Let us take up the case when $x = 0$.

From (6.88) we expand $\bar{\phi}(s)$ as follows

$$\bar{\phi}(s) = \phi_0 \left(\frac{1}{s} + \frac{1}{\alpha^2 h^2} + \frac{1}{\alpha^4 h^4} + \ldots \right) - \phi_0 s^{-\frac{1}{2}} \left(\frac{1}{\alpha h} + \frac{s}{\alpha^3 h^3} + \frac{s^2}{\alpha^5 h^5} \right) \quad (6.89)$$

By the asymptotic inversion formula of Laplace transform given in $(C8)$ of Appendix C we have for large t

$$\phi(t) = \phi_0 - \phi_0 \frac{\sin \frac{\pi}{2}}{\pi} \left[\frac{\Gamma(\frac{1}{2})}{\alpha h \sqrt{t}} - \frac{1}{\alpha^3 h^3} \frac{\Gamma(\frac{3}{2})}{t^{\frac{3}{2}}} + \frac{1}{\alpha^5 h^5} \frac{\Gamma(\frac{5}{2})}{t^{\frac{5}{2}}} - \ldots \right] \quad (6.90)$$

where we have set $\beta = \frac{1}{2}$. A more compact form is

$$\phi(t) = \phi_0 - \frac{\phi_0}{\sqrt{\pi}} \left[\frac{1}{\alpha h \sqrt{t}} - \frac{1}{\alpha^3 h^3 t^{\frac{3}{2}}} + \frac{1 \cdot 3}{4\alpha^5 h^4 t^{\frac{5}{2}}} - \ldots \right] \quad (6.91)$$

giving the asymptotic nature of $\phi(t)$.

The case $x \neq 0$ where the exponential in (6.88) contributes some additional terms is handled similarly and is left as an exercise.

Example 6.8

Find the asymptotic expansion of $\phi(t)$ for when $\phi(t)$ satisfies the heat equation

$$\phi_t(x, t) = \alpha^2 \phi_{xx}(x, t), \quad x \geq 0, \quad t > 0 \quad (6.92)$$

subject to the initial condition

$$\phi(x, 0) = 0, \quad x \geq 0 \quad (6.93)$$

and an oscillatory fluctuation at $x = 0$

$$\phi(0, t) = a \cos(\omega t), \quad t > 0 \quad (6.94)$$

where a is a constant and ω is the frequency of oscillation.

Taking the Laplace transform of both sides of (6.92) yields the form similar to (6.83) in which (6.94) is used. Its bounded solution is given by (6.85).

Taking now the Laplace transform of the boundary condition (6.94) we find

$$\bar{\phi}(0, s) = \frac{as}{s^2 + \omega^2} \quad (6.95)$$

where we have employed the Laplace transform of the cosine function from Table $C1$.

The bounded solution (6.85) when confronted with (6.95) yields the following result

$$\bar{\phi}(x, s) = \frac{as}{s^2 + \omega^2} e^{-x\sqrt{\frac{s}{\alpha}}} \tag{6.96}$$

Notice that the singularities of $\bar{\phi}(x, s)$ are the simple poles located at $s = \pm i\omega$ and the branch point $s = 0$. Writing

$$\phi(t) = \phi_+(t) + \phi_-(t) + \phi_0(t) \tag{6.97}$$

where $+, -$ and 0 denote the contributions from the three singularities, we have specifically for $\phi_+(t)$

$$\phi_+(t) = \text{residue of} \quad e^{st} \frac{as}{s^2 + \omega^2} e^{-x\sqrt{\frac{s}{\alpha}}} \quad \text{at} \quad s = i\omega \tag{6.98}$$

The result is

$$\phi_+(t) = e^{i\omega t} \left(\frac{a \cdot 2i\omega}{2i\omega}\right) e^{-x\sqrt{\frac{\omega}{\alpha}} e^{i\frac{\pi}{4}}} = \frac{a}{2} e^{i(\omega t - x\sqrt{\frac{\omega}{2\alpha}})} e^{-x\sqrt{\frac{\omega}{2\alpha}}} \tag{6.99}$$

Similarly $\phi_-(t)$ is given by

$$\phi_-(t) = \frac{a}{2} e^{-i(\omega t - x\sqrt{\frac{\omega}{2\alpha}})} e^{-x\sqrt{\frac{\omega}{2\alpha}}} \tag{6.100}$$

To estimate $\phi_0(t)$ the best we can do is to go for an asymptotic evaluation. Towards this end we note the following expansion of the right side of (6.96) around $s = 0$

$$\frac{as}{s^2 + \omega^2} e^{-x\sqrt{\frac{s}{\alpha}}} = \frac{a}{\omega^2}\left(s + \frac{s^2 x^2}{2\alpha} + ...\right) - \frac{a}{\omega^2} s^{-\frac{1}{2}}\left(\frac{xs^2}{\alpha} + \frac{x^3 s^3}{6\alpha^3} + ...\right) \tag{6.101}$$

This shows from the asymptotic inversion formula

$$\phi_0(t) \sim -\frac{a}{\omega^2} \frac{\sin\frac{\pi}{2}}{\pi}\left[\frac{x}{\alpha} \frac{\Gamma(\frac{5}{2})}{t^{\frac{5}{2}}} - \frac{x^3}{6\alpha^3} \frac{\Gamma(\frac{7}{2})}{t^{\frac{7}{2}}} + ...\right] \tag{6.102}$$

Combining (6.99), (6.100) and (6.102) we arrive at the result

$$\phi(t) \sim a e^{-x\sqrt{\frac{\omega}{2\alpha}}} \cos(\omega t - x\sqrt{\frac{\omega}{2\alpha}}) - \frac{a}{\pi\omega^2}\left[\frac{x}{\alpha} \frac{\Gamma(\frac{5}{2})}{t^{\frac{5}{2}}} - \frac{x^3}{6\alpha^3} \frac{\Gamma(\frac{7}{2})}{t^{\frac{7}{2}}} + ...\right] \tag{6.103}$$

which stands for the asymptotic form for $\phi(t)$.

6.3 Summary

The general aim in this chapter was to solve some typical problems of PDEs by the use of integral transform methods as a convenient mathematical tool. We however concentrated only on the Fourier and Laplace transforms although one should note that other transforms could also be used as a means to generate solutions of a PDE. We made use of their properties of the transforms collected in Appendix B and Appendix C. We also considered a couple of examples for which the asymptotic form of the solution could be derived. Such asymptotic forms are often found useful in many classes of problems holding practical interest.

Exercises

1. Find the Fourier integral representation of the single pulse function

$$f(x) = \begin{cases} 1 & \text{if } |x| < 1 \\ 0 & \text{if } |x| > 1 \end{cases}$$

Deduce the Dirichlet integral

$$\int_0^\infty \frac{\sin p}{p} \, dp = \frac{\pi}{2}$$

2. Find by using Fourier transform the fundamental solution of the PDE

$$\phi_t = \phi_{xx} - x\phi_x$$

as $\phi(x,t) = Ce^t e^{\frac{x^2}{2}}$, where C is a constant.

3. Small amplitude water waves produced by an inertial surface displacement are controlled by a two-dimensional Laplace's equation

$$\phi_{xx} + \phi_{zz} = 0, \quad -\infty < z < \eta$$

where x and z are respectively the horizontal and upward vertical axis, $\phi(x,z,t)$ is the velocity potential and $\eta(x,t)$ is the elevation of the free surface subject to the initial conditions

$$\phi(x,z,0) = 0, \quad \eta(x,0) = f(x),$$

where $f(x)$ is an even function of x, the boundary conditions

$$\eta_t - \phi_z = 0, \quad \phi_t + g\eta = 0 \quad \text{both at} \quad z = 0$$

and the smoothness condition

$$\phi(x,z,t) \to 0 \quad \text{as} \quad z \to -\infty$$

Use the method of Fourier transform, find an integral expression of $\eta(x,t)$ in the form

$$\eta(x,t) = \frac{1}{\sqrt{2\pi}} \int_0^\infty f(p)[\cos(px - \sqrt{gp}\,t) + \cos(px + \sqrt{gp}\,t)]$$

where p is the transform parameter.

4. Consider a three-dimensional Laplace's equation

$$\phi_{xx} + \phi_{yy} + \phi_{zz} = 0, \quad -\infty < x < \infty, \quad z \geq 0$$

subject to the boundary condition

$$\phi(x, y, 0) = f(x, y)$$

and the smoothness conditions

$$\phi(x, y, z) \quad \to \quad 0 \quad \text{as} \quad z \quad \to \quad \infty \quad \text{and also as} \quad x, y \quad \to \quad \pm\infty$$

along with

$$\phi_x, \phi_y \quad \to \quad 0 \quad \text{as} \quad x, y \quad \to \quad \pm\infty$$

Find the solution in the form

$$\phi(x, y, z) = \frac{z}{2\pi} \int_{-\infty}^{+\infty} \int_{-\infty}^{+\infty} \frac{f(\xi, \eta)}{[(x - \xi)^2 + (y - \eta)^2 + z^2]^{\frac{3}{2}}} d\xi d\eta$$

5. Solve the following generalized form of the telegraph equation by the Fourier transform method

$$\phi_{tt} + (\alpha + \beta)\phi_t + \alpha\beta\phi = c^2\phi_{xx}$$

where α and β are non-zero constants and show that the solution reveals dispersion. Discuss the case when $\alpha = \beta$.

6. Use Laplace transform to write down the solution of the PDE

$$\phi_{xt} + x\phi_x + 2\phi = 0$$

subject to the initial-boundary conditions

$$\phi(x, 0) = 1, \quad \phi(0, t) = e^{-at}, \quad a > 0$$

in the form

$$\phi(x, t) = \frac{1}{2i\pi} \int_{r-i\infty}^{r+i\infty} e^{st}\zeta(s)ds, \quad s > a$$

where $\zeta(s) = \frac{s^2}{s(s+a)(x+s)^2}$.

7. Consider the problem of a semi-infinite string described by the PDE

$$\phi_{tt} = c^2\phi_{xx} + \phi_0, \quad 0 < x < \infty, \quad t > 0$$

where one end of the string is held fixed and ϕ_0 is the constant external force acting on it. Further, the string is subjected to the initial boundary conditions

$$\phi(x,0) = 0, \quad \phi_t(x,0) = 0, \quad \phi(0,t) = 0$$

Assume that $\phi_x(x,t) \to 0$ as $x \to \infty$. Solve by the Laplace transform method to derive $\phi(x,t)$ in the form

$$\phi(x,t) = \begin{cases} \frac{\phi_0}{2}[t^2 - (t - \frac{x}{c})^2] & \text{if } t \geq \frac{x}{c} \\ \frac{\phi_0}{2}t^2 & \text{if } t \leq \frac{x}{c} \end{cases}$$

8. Solve the wave equation

$$\phi_{tt} = c^2 \phi_{xx}, \quad 0 < x < \infty, \quad t > 0$$

which is subjected to the initial boundary conditions

$$\phi(x,0) = 0, \quad \phi_t(x,0) = 0, \quad \phi(0,t) = f(t)$$

where $f(t)$ is a given function. Assume that $\phi_x(x,t) \to 0$ as $x \to \infty$. Solve by the Laplace transform method to derive $\phi(x,t)$ in the form

$$\phi(x,t) = \begin{cases} \sin(t - \frac{x}{c}) & \text{if } \frac{x}{c} < t < \frac{x}{c} + 2\pi \\ 0 & \text{otherwise} \end{cases}$$

9. Solve the following fourth order PDE

$$\phi_{tt} = \lambda^2 \phi_{xxxx}, \quad 0 < x < l, \quad t > 0$$

which is subjected to the initial conditions

$$\phi(x,0) = 0, \quad \phi_t(x,0) = \phi_0 \sin(\frac{\pi x}{l})$$

and boundary conditions

$$\phi(0,t) = 0, \quad \phi(l,t) = 0, \quad \phi_{xx}(0,t) = 0. \quad \phi_{xx}(l,t) = 0$$

where λ is a constant.

10. Focus on the homogeneous heat equation

$$\phi_t(x,t) = \phi_{xx}(x,t), \quad -\infty < x < \infty, \quad t > 0$$

subjected to the following initial data on time at $t = s$

$$\phi(x, s; s) = f(x, s), \quad -\infty < x < \infty$$

In terms of the Green's function $G(x, s) = \frac{1}{\sqrt{4\pi t}} e^{-\frac{x^2}{4t}}$ the solution of the problem is given by

$$\phi(x, t; s) = \int_{-\infty}^{\infty} G(x - \xi, t - s) f(\xi, s) d\xi, \quad -\infty < x < \infty$$

Establish Duhamel's principle which states that in an infinite domain the solution of the corresponding inhomogeneous problem namely,

$$\phi_t(x, t) = \phi_{xx}(x, t) + f(t, x), \quad -\infty < x < \infty, \quad t > 0$$

subject to the vanishing initial condition $\phi(x, 0) = 0$ is given by

$$\phi(x, t) = \int_0^t \phi(x, t; s) ds$$

i.e. $\phi(x, t)$ has the form

$$\phi(x, t) = \int_{s=0}^t \int_{\xi=-\infty}^{\infty} G(x - \xi, t - s) f(\xi, s) \phi(x, t; s) ds$$

which translates to

$$\phi(x, t) = \int_{s=0}^t \int_{\xi=-\infty}^{\infty} \frac{1}{\sqrt{4\pi(t - s)}} e^{-\frac{(x-\xi)^2}{4(t-s)}} f(\xi, s) d\xi ds$$

A problem along similar lines can be set up and solved for the heat equation defined over a finite interval. Duhamel's principle also holds for the solvability of the inhomogeneous wave equation where the source term is transferred to the initial velocity for the corresponding homogeneous problem.

Appendix A

Dirac delta function

Dirac delta function $\delta(x)$

The delta function was introduced by P.A.M. Dirac in the development of the theory of quantum mechanics in his 1930 book *The Principles of Quantum Mechanics* and hence it goes by his name. His point was that, as he wrote, $\delta(x)$ *is not a quantity which can be generally used in mathematical analysis like an ordinary function, but its use must be confined to certain simple types of expression for which it is obvious that no inconsistency can arise.* However, the history[1] of delta function is far older[2] and dates back to the work of Gustav Kirchhoff who studied it in 1882 as the limit of the functions

$$\delta_n(x) = \sqrt{\frac{n}{\pi}} e^{-nx^2}, \quad n = 1, 2, 3, \ldots$$

Delta function was also used a decade later in 1892 by the electrical engineer Oliver Heaviside in a paper entitled *On operations in physical mathematics*. Later, in a wider setting, the famous Russian mathematician Sergei Lvovich Sobolev introduced the concept of distribution theory in 1935 and subsequently the French mathematician Laurent Schwartz gave an elaborate exposition of it in the two volumes of the book entitled *Theory of Distributions* appearing in 1950 and 1951. In the literature a distribution is also referred to as a generalized function. The delta function is the analog of the piecewise function Kronecker delta for the continuous case. The latter is defined for two indices i and j by $\delta_{ij} = 1, i = j$ but $= 0, i \neq j$. Both for the delta function and Kronecker delta the same symbol δ is retained.

The Dirac delta function, denoted by $\delta(x)$, is a mathematical object equipped with the property

---

[1] J. Lighthill, *Fourier Analysis and Generalized functions*, Tributes to Paul Dirac, Ed. J.G.Taylor, Adam Hilger, IOP Publishing, Bristol, 1987.

[2] Although no in-depth study was made, the presence of delta function was noted by Poisson in 1815, Fourier in 1822 and Cauchy in 1823 and 1827.

$$\int_{-\infty}^{+\infty} f(x)\delta(x)dx = f(0) \tag{A.1}$$

for any continuous function $f(x)$. Furthermore, $\delta(x)$ is vanishing everywhere except at $x = 0$

$$\delta(x) = 0 \quad \text{for all} \quad x \neq 0 \tag{A.2}$$

On the other hand, setting $f(x) = 1$ implies that the integral of the delta function over $(-\infty, +\infty)$ is unity

$$\int_{-\infty}^{+\infty} \delta(x)dx = 1 \tag{A.3}$$

(A.1), (A.2) and (A.3) are the essential characteristics that distinguish a delta function from an ordinary function.

Instead of $x = 0$ if the argument of the delta function is at the point $x = \xi$ then the above properties take the forms

$$\delta(x - \xi) = 0, \quad x \neq \xi \tag{A.4}$$

$$\int_a^b \delta(x - \xi)dx = 1, \quad a \leq \xi \leq b, \quad \int_a^b \delta(x - \xi)dx = 0, \quad a, b < \xi \quad \text{and} \quad a, b > \xi \tag{A.5}$$

$$\int_{-\infty}^{+\infty} f(x)\delta(x - \xi)dx = f(\xi) \tag{A.6}$$

where if $f(x) = 1$ then

$$\int_{-\infty}^{+\infty} \delta(x - \xi)dx = 1 \tag{A.7}$$

We refer to (A.1) or (A.6) as the reproductive property of the delta function. It is also called the sifting property or selector property.

If for a sequence of strongly peaked classical functions $g_k(x)$ such that as $k \to \infty$ the following integral character is revealed

$$\lim_{k \to \infty} \int_{-\infty}^{+\infty} g_k(x)f(x)dx = f(0), \quad k = 1, 2, ...$$

then such a sequence $g_k(x), k = 1, 2, ...$ is called a delta sequence and the delta function may be viewed as a limit form of it.

The delta function has a relevance to multi-dimensional problems too. Suppose we are dealing with a set of spherical polar coordinates r, θ, ϕ. The Jacobian matrix which is defined by

$$J = \begin{pmatrix} \frac{\partial x}{\partial r} & \frac{\partial x}{\partial \theta} & \frac{\partial x}{\partial \phi} \\ \frac{\partial y}{\partial r} & \frac{\partial y}{\partial \theta} & \frac{\partial y}{\partial \phi} \\ \frac{\partial z}{\partial r} & \frac{\partial z}{\partial \theta} & \frac{\partial z}{\partial \phi} \end{pmatrix} \tag{A.8}$$

becomes for $x = r \sin\theta \cos\phi, y = r \sin\theta \sin\phi, z = r \cos\theta$

$$J = \begin{pmatrix} \sin\theta \cos\phi & r \cos\theta \cos\phi & -r \sin\theta \sin\phi \\ \sin\theta \sin\phi & r \cos\theta \sin\phi & r \sin\theta \cos\phi \\ \cos\theta & -r \sin\theta & 0 \end{pmatrix} \tag{A.9}$$

The value of the determinant of the matrix is $r^2 \sin\theta$. Hence the delta function in spherical polar coordinates reads

$$(r^2 \sin\theta)^{-1} \delta(r - r') \delta(\theta - \theta') \delta(\phi - \phi') \tag{A.10}$$

In two dimensions since the Jacobian is $J = r$, the delta function is

$$(r)^{-1} \delta(r - r') \delta(\theta - \theta') \tag{A.11}$$

In cylindrical coordinates $x = r \cos\phi, y = r \sin\phi, z = z$, making appropriate changes in variables in (A.8), the Jacobian is easily worked out to be simply $J = r$. Hence the delta function in cylindrical coordinates reads

$$(r)^{-1} \delta(r - r') \delta(\phi - \phi') \delta(z - z') \tag{A.12}$$

(A.10) and (A.12) are to be compared with the Cartesian situation where it is $\delta(x - x') \delta(y - y') \delta(z - z')$.

We now look for sequence of functions, the delta sequence, as mentioned a little earlier, that satisfies the reproductive property or sifting property. One such sequence is provided by the Dirichlet's formula

$$\lim_{k \to \infty} \int_{-\infty}^{+\infty} f(x) \frac{\sin(kx)}{\pi x} dx = f(0), \quad k = 1, 2, \ldots \tag{A.13}$$

where $f(x)$ is a bounded and differentiable function.

The representation of rectangular blocks defined as the limit

$$\lim_{k \to \infty} \int_{-\infty}^{+\infty} f(x) u_k(x) dx = f(0), \quad k = 1, 2, \ldots \tag{A.14}$$

of a family of functions $u_k(x)(x)$

$$u_k(x) = \begin{cases} \frac{k}{2} & \text{if } |x| < \frac{1}{k} \\ 0 & \text{otherwise,} \end{cases}$$

is another example.

One can also visualize the limit of the Gaussian

$$\lim_{k\to\infty}\int_{-\infty}^{+\infty} f(x)\xi_k(x)dx = f(0), \quad k = 1, 2, \dots \tag{A.15}$$

where $\xi_k(x)$ is (see Figure A.1)

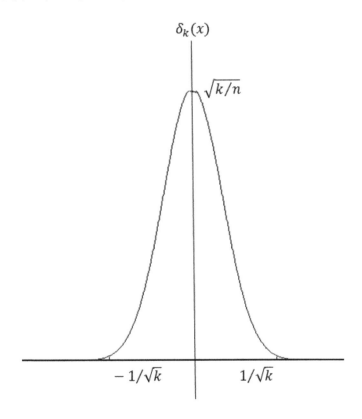

FIGURE A.1: Gaussian profile.

$$\xi_k(x) = \sqrt{\frac{k}{\pi}}e^{-kx^2}, \quad k = 1, 2, \dots \tag{A.16}$$

and the so-called Lorentzian

$$\lim_{k\to\infty}\int_{-\infty}^{+\infty} f(x)\zeta_k(x)dx = f(0), \quad k = 1, 2, \dots \tag{A.17}$$

where $\zeta_k(x)$ is (see Figure A.2)

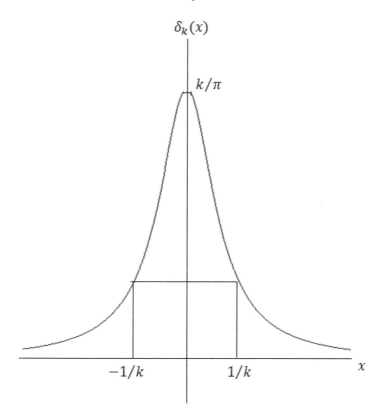

FIGURE A.2: Lorentzian profile.

$$\zeta_k(x) = \frac{1}{\pi} \frac{k}{1 + k^2 x^2}, \quad k = 1, 2, \ldots \tag{A.18}$$

as legitimate candidates for the delta sequence. We also note some of the other choices which include the following functions

$$\frac{1}{2\epsilon} exp(-\frac{|x|}{\epsilon}), \quad \frac{1}{2\sqrt{\pi\epsilon}} exp(-\frac{x^2}{4\epsilon}), \quad \frac{1}{2\epsilon} \frac{1}{\cosh^2(\frac{x}{\epsilon})} \tag{A.19}$$

which in the limit $\epsilon \to 0$ reflect the standard characteristics of a delta function.

Note that the Gaussian and Lorentzian can also be respectively translated to the form

$$\lim_{\alpha \to 0} \frac{1}{\sqrt{2\pi}\alpha} e^{-\frac{x^2}{2\alpha^2}} \tag{A.20}$$

These can be also looked upon as the limit of the Cauchy distribution

$$\lim_{\epsilon \to 0} \frac{1}{\pi} \frac{\epsilon}{x^2 + \epsilon^2} \tag{A.21}$$

The sequence of functions $\frac{\sin(kx)}{\pi x}$ or $u_k(x)$ or $\xi(x)$ or $\zeta(x)$, $k = 1, 2, ...$, is a representative of a delta sequence.

Let us attend to $\frac{\sin(kx)}{\pi x}$ first. The following integral values are well known

$$\int_0^{+\infty} \frac{\sin(x)}{\pi x} dx = \frac{1}{2} \tag{A.22}$$

and

$$\int_{-\infty}^0 \frac{\sin(x)}{\pi x} dx = \frac{1}{2} \tag{A.23}$$

When put together we have

$$\int_{-\infty}^{+\infty} \frac{\sin(x)}{\pi x} dx = 1 \tag{A.24}$$

and a change of variable allows us to write

$$\int_{-\infty}^{+\infty} \frac{\sin(kx)}{\pi x} dx = 1 \tag{A.25}$$

We now look at the finite range of integral over (a, b) where $b > a > 0$. Then the parameter k can be easily transferred to the limits of the integral

$$\frac{1}{\pi} \int_a^b \frac{\sin(kx)}{x} dx = \frac{1}{\pi} \int_{ak}^{bk} \frac{\sin(t)}{t} dt, \quad b > a > 0 \tag{A.26}$$

where the right side $\to 0$ as $k \to \infty$. Likewise

$$\frac{1}{\pi} \int_{ak}^{bk} \frac{\sin(t)}{t} dt \quad \to 0 \quad \text{as} \quad k \to \infty, \qquad 0 < a < b \tag{A.27}$$

Next, due to the above vanishing results on either side of zero, we have for any $\epsilon > 0$, however small

$$\lim_{k \to \infty} \int_{-\infty}^{+\infty} f(x) \frac{\sin(kx)}{\pi x} dx = \lim_{k \to \infty} \int_{-\epsilon}^{+\epsilon} f(x) \frac{\sin(kx)}{\pi x} dx \tag{A.28}$$

where the right side can be put as

$$f(0) \lim_{k \to \infty} \int_{-\epsilon}^{+\epsilon} \frac{\sin(kx)}{\pi x} dx \tag{A.29}$$

with the help of the mean value theorem of integral calculus.

In (A.29) we can, without any loss of generality, extend the range of integration to $(-\infty, +\infty)$ resulting in

$$f(0) \lim_{k \to \infty} \int_{-\infty}^{+\infty} \frac{\sin(kx)}{\pi x} dx \qquad (A.30)$$

which when (A.25) is used boils down to simply $f(0)$ and we conclude

$$\lim_{k \to \infty} \int_{-\infty}^{+\infty} f(x) \frac{\sin(kx)}{\pi x} dx = f(0) \qquad (A.31)$$

We thus recover the Dirichlet's formula signalling that in the limit $k \to \infty$, $\frac{\sin(kx)}{\pi x}$ is a representation of the Dirac delta function

$$\lim_{k \to \infty} \frac{\sin(kx)}{\pi x} = \delta(x) \qquad (A.32)$$

Equivalently,

$$\lim_{\epsilon \to 0^+} \frac{\sin(\frac{x}{\epsilon})}{\pi x} = \delta(x) \qquad (A.33)$$

Further since

$$\frac{1}{2\pi} \int_{-k}^{+k} e^{irx} dr = \frac{\sin(kx)}{\pi x} \qquad (A.34)$$

we also have another representation of the delta function

$$\frac{1}{2\pi} \int_{-\infty}^{+\infty} e^{ikx} dk = \delta(x) \qquad (A.35)$$

We now turn to the behaviour of $u_k(x)$. Given its definition we find

$$\lim_{k \to \infty} \int_{-\infty}^{+\infty} u_k(x) f(x) dx = \frac{k}{2} \int_{-\frac{1}{k}}^{+\frac{1}{k}} f(x) dx \qquad (A.36)$$

where the integral in the right side, by the mean value theorem of integral calculus is $\frac{2}{k} f(\xi)$, where $|\xi| < \frac{1}{k}$, and as such

$$\lim_{k \to \infty} \int_{-\infty}^{+\infty} u_k(x) f(x) dx = f(\xi) \qquad (A.37)$$

But as $k \to \infty$ the right side goes to $f(0)$ and hence we can write

$$\lim_{k \to \infty} \int_{-\infty}^{+\infty} u_k(x) f(x) dx \to f(0) \qquad (A.38)$$

which reflects the property of a delta sequence.

In a similar way if the argument of u_k is different from zero, then a change of variable gives

$$\lim_{k\to\infty} \int_{-\infty}^{+\infty} u_k(x-x_0)f(x)dx = \lim_{k\to\infty} \int_{-\infty}^{+\infty} u_k(y)f(y+x_0)dy \to f(x_0) \qquad \text{(A.39)}$$

where we employed the mean value theorem of integral calculus to express $\int_{-\infty}^{+\infty} u_k(y)f(y+x_0)dy = \frac{k}{2}\frac{2}{k}f(\xi+x_0)$, where $|\xi| < \frac{1}{k}$ and exploited the limit $k \to \infty$. In essence this is the reproductive or sifting property of the delta function projecting out the value of f at $x = x_0$.

If the argument of the delta function involves a scaling $x \to \eta x$ then we have for $\eta > 0$

$$\lim_{k\to\infty} \int_{-\infty}^{+\infty} u_k(\eta x)f(x)dx = \lim_{k\to\infty} \int_{-\infty}^{+\infty} u_k(y)f(\frac{y}{\eta})\frac{dy}{\eta} \to \frac{1}{\eta}f(0) \qquad \text{(A.40)}$$

while for $\eta < 0$

$$\lim_{k\to\infty} \int_{-\infty}^{+\infty} u_k(\eta x)f(x)dx = -\lim_{k\to\infty} \int_{-\infty}^{+\infty} u_k(y)f(\frac{y}{\eta})\frac{dy}{\eta} \to -\frac{1}{\eta}f(0) \qquad \text{(A.41)}$$

The above two results when united furnish an important result of scaling

$$\delta(\eta x) = \frac{1}{|\eta|}\delta(x) \qquad \text{(A.42)}$$

Next, we look at the Gaussian $\xi_k(x)$. We first of all transform the integral

$$\int_{-\infty}^{+\infty} e^{-x^2}dx = \sqrt{\pi} \qquad \text{(A.43)}$$

to the form

$$\int_{-\infty}^{+\infty} e^{-kx^2}dx = \sqrt{\frac{\pi}{k}} \qquad \text{(A.44)}$$

on replacing x by $x\sqrt{k}$. This means

$$\int_{-\infty}^{+\infty} \xi_k(x)dx = 1, \quad k = 1, 2, ... \qquad \text{(A.45)}$$

Let us consider at the finite range of integration over (a,b), where $b > a > 0$. Then

$$\int_a^b \xi_k(x)dx = \frac{1}{\sqrt{\pi}} \int_{a\sqrt{k}}^{b\sqrt{k}} e^{-y^2}dy \to 0 \quad \text{as} \quad k \to \infty \qquad \text{(A.46)}$$

The above limiting result also holds for $a < b < 0$.

Decomposing $(-\infty, +\infty)$ to pieces $(-\infty, -\epsilon)$, $(-\epsilon, +\epsilon)$ and $(+\epsilon, +\infty)$, $\epsilon > 0$ no matter how small, and making use of the above results we find that we can write

$$\lim_{k\to\infty} \int_{-\infty}^{+\infty} \xi_k(x)f(x)dx = \lim_{k\to\infty} \int_{=\epsilon}^{+\epsilon} \xi_k(x)f(x)dx \tag{A.47}$$

where the right side by the use of mean value theorem of integral calculus can be written as

$$f(0)\lim_{k\to\infty} \int_{=\epsilon}^{+\epsilon} \xi_k(x)dx \tag{A.48}$$

Stretching the integral domain to $(-\infty, +\infty)$, which as shown earlier has the value equal to unity, we arrive at the desired reproductive result

$$\lim_{k\to\infty} \int_{=\infty}^{+\infty} \xi_k(x)f(x)dx = f(0) \tag{A.49}$$

Finally we deal with the Lorentzian case. If we set $g(x) = f(x) - f(0)$ which implies $g(0) = 0$, then to establish the reproductive property

$$\lim_{k\to\infty} \int_{=\infty}^{+\infty} \zeta_k(x)f(x)dx = f(0) \tag{A.50}$$

it suffices if we can show

$$\lim_{k\to\infty} \int_{=\infty}^{+\infty} \zeta_k(x)g(x)dx = 0 \tag{A.51}$$

Proceeding towards this aim, we choose some appropriate number $a > 0$ and decompose $(-\infty, +\infty)$ around $(-a, +a)$ to write

$$\int_{=\infty}^{+\infty} \zeta_k(x)g(x)dx = \int_{=\infty}^{-a} \zeta_k(x)g(x)dx + \int_{=a}^{+a} \zeta_k(x)g(x)dx + \int_{+a}^{+\infty} \zeta_k(x)g(x)dx \tag{A.52}$$

Concerning the integral over $(-a, +a)$, if $|g(x)| \le \mu(a)$ there, $\mu(a)$ being the maximum value, then

$$\left| \int_{=a}^{+a} \zeta_k(x)g(x)dx \right| \le \int_{=a}^{+a} \zeta_k(x)|g(x)|dx \le \mu(a)\int_{=a}^{+a} \zeta_k(x)dx \tag{A.53}$$

where the definite integral in the right side has the value $\frac{2}{\pi}\tan^{-1}(ak)$, it follows that

$$\left| \int_{=a}^{+a} \zeta_k(x)g(x)dx \right| \le \mu(a) \tag{A.54}$$

Since by definition of $g(x)$, $g(x) = 0$ we can claim that $\lim_{a \to 0} \mu(a) = 0$ and write

$$|\int_{=a}^{+a} \zeta_k(x) g(x) dx| < \frac{\epsilon}{2} \qquad (A.55)$$

where ϵ is a quantity no matter how small.

For the other two integrals we have

$$|\int_{=\infty}^{-a} \zeta_k(x) g(x) dx + \int_{+a}^{+\infty} \zeta_k(x) g(x) dx| \leq \nu [\int_{=\infty}^{-a} \zeta_k(x) dx + \int_{+a}^{+\infty} \zeta_k(x) dx] \qquad (A.56)$$

where since $f(x)$ is bounded in $(-\infty, +\infty)$ and $g(0) = 0$, $g(x)$ is bounded too in $(-\infty, +\infty)$ and given by $|g(x)| \leq \nu$. After the easy evaluation of the integrals, the right side has the value $\nu(1 - \frac{2}{\pi} \tan^{-1}(ak))$ which $\to 0$ as $k \to \infty$. Thus we can write

$$|\int_{=\infty}^{-a} \zeta_k(x) g(x) dx + \int_{+a}^{+\infty} \zeta_k(x) g(x) dx| \leq \frac{\epsilon}{2} \qquad (A.57)$$

in terms of the small quantity ϵ.

All this means that we have

$$\int_{=\infty}^{+\infty} \zeta_k(x) g(x) dx \leq \frac{\epsilon}{2} + \frac{\epsilon}{2} = \epsilon \qquad (A.58)$$

This being so the reproductive property holds.

Other results using delta function

The following results often prove to be useful:

$$\begin{aligned} (i) \quad & x\delta(x) & = & \quad 0 \\ (ii) \quad & \delta(x) & = & \quad \delta(-x) \end{aligned}$$

$$(iii) \quad \int \delta(x - \xi)\delta(x - \eta) dx \quad = \quad \delta(\xi - \eta) \qquad (A.59)$$

The proof of the first result is trivial. For the second result since we can write

$$\delta(-x) = \frac{1}{2\pi} \int_{-\infty}^{+\infty} e^{-ikx} dk \qquad (A.60)$$

replacing k by $-k$ in the right side and changing the limits of the integral accordingly a representation of $\delta(x)$ easily follows and shows that $\delta(x) = \delta(-x)$ implying that the delta function is an even function.

To prove (iii) let us represent the integral on the left side as

$$\int dx \delta(x - \xi)\delta(x - \eta) = \frac{1}{(2\pi)^2} \int dx \int_{-\infty}^{+\infty} e^{iy(x-\xi)} dy \int_{-\infty}^{+\infty} e^{it(x-\eta)} dt \qquad (A.61)$$

where the right side can be arranged to the following form

$$\frac{1}{2\pi} \int dy \int dt (\frac{1}{2\pi} \int e^{i(y+t)x} dx) e^{-i\xi y - i\eta t} \qquad (A.62)$$

It is clear that the integral over x is simply $\delta(y + t)$. So if we perform the integral over t we will be left with

$$\frac{1}{2\pi} \int e^{-i(\xi-\eta)y} dy \qquad (A.63)$$

This integral, on using the property that the delta function is an even function, stands for $\delta(\xi - \eta)$ and we have justified the required assertion.

We often have to face situations when the argument of the delta function represents a function such as $\delta(f(x))$. Setting aside complicated cases let us look at a monotonic function which possesses a simple zero at $x = \xi$. If $f'(\xi) > 0$ we can express for a smooth function $g(x)$

$$\int_{-\infty}^{+\infty} \delta(f(x))g(x)dx = \int_{-\infty}^{+\infty} \delta(y)g(x(y))\frac{dy}{f'(x)} = \frac{g(\xi)}{f'(\xi)} \qquad (A.64)$$

where we have replaced $f(x)$ by the variable y and noted that the vanishing of the argument of $\delta(y)$ takes place for the corresponding $x = \xi$ value. Since the right side can also be expressed in the form

$$\int_{-\infty}^{+\infty} \frac{\delta(x - \xi)g(x)}{f'(\xi)} dx \qquad (A.65)$$

it therefore follows that

$$\delta(f(x)) = \frac{\delta(x - \xi)}{f'(\xi)} dx \qquad (A.66)$$

On the other hand, if $f'(\xi) < 0$ we encounter a sign change only and so combining both the cases we have the result

$$\delta(f(x)) = \frac{\delta(x - \xi)}{|f'(\xi)|} dx \qquad (A.67)$$

When $f(x)$ runs into multiple zeros located at $(\xi_1, \xi_2, , \xi_n)$ we get the general form

$$\delta(f(x)) = \sum_{k=1}^{n} \frac{\delta(x - \xi_k)}{|f'(\xi_k)|} dx \qquad (A.68)$$

As applications of the above formula it is easy to show

$$(i) \quad \delta(x^2 - \xi^2) \quad = \quad \frac{1}{2\xi}[\delta(x-\xi) + \delta(x+\xi)]$$

$$(ii) \quad \delta(\tan x) \quad = \quad \delta(x), \quad x \in \left(-\frac{\pi}{2}, \frac{\pi}{2}\right)$$

$$(iii) \quad \delta(\cos x) \quad = \quad \delta\left(x - \frac{\pi}{2}\right), \quad 0 < x < \pi$$

$$(iv) \quad \int_{-2\pi}^{+2\pi} e^{\pi x}\delta(x^2 - \pi^2) \quad = \quad \frac{1}{\pi}\cosh(\pi^2) \tag{A.69}$$

Test function

Consider a set S of real numbers. Its limit point α is such that every neighbourhood of it, no matter how small, contains at least one point of S. The set S with all its limit points constitutes the closure of S and denoted by $\bar{S}$.

The support of a function addresses the smallest closed set where it takes on nonzero values but vanishes identically outside of it. The support of a function f is denoted by supp(f). A compact support is the one in which the supp(f) is compact i.e. it is a bounded set.

The example of $f(x) = \sin(x)$ is an interesting one in the sense that its support runs over the entire real axis although $\sin(x)$ vanishes at all points $x = n\pi, n = 1, 2, ...$ Another example is (see Figure A.3)

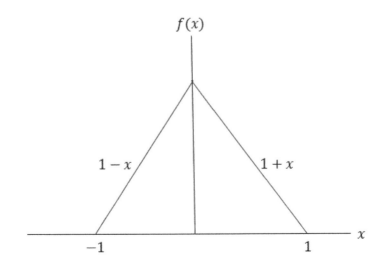

FIGURE A.3: The function f(x).

$$f(x) = \begin{cases} 0, & \text{if} \quad -\infty < x \leq -1 \\ 1+x, & \text{if} \quad -1 < x < 0 \\ 1-x, & \text{if} \quad 0 \leq x < 1 \\ 0, & \text{if} \quad 1 \leq x < \infty \end{cases}$$

It is clear that the function $f(x)$ is different from zero in the open interval $(-1, 1)$, the closure of the latter is $[-1, 1]$ and so $[-1, 1]$ stands for the compact support of $f(x)$.

With these introductory remarks we are in a position to define a test function. Real valued functions denoted by $\phi(x) = \phi(x_1, x_2, ..., x_n)$ whose derivatives of all orders exist (in other words, they are infinitely differentiable at all points in $\Re$) and have a compact support in a closed bounded subset of $\Re$ are called test functions. Test functions are smooth, with no sharp features in their graphs. An explicit illustration of a test function is (see Figure A.4)

$$\phi(x) = \begin{cases} e^{-\frac{1}{1-|x|^2}} & \text{if} \quad |x| < 1 \\ 0 & \text{otherwise,} \end{cases}$$

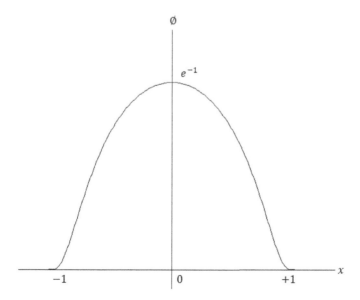

FIGURE A.4: Test function $\phi(x)$.

It gives a compactly supported test function $\phi(x)$ which is differentiable to any order. Note, however, that there exist numerous classes of smooth functions (for example, a polynomial or an exponential) which do not have a compact support. The space is denoted by D and addresses the family of infinitely differentiable compact functions.

As another example consider the function $g(x)$ defined by

$$g(x) = \begin{cases} exp(-\frac{1}{x}) & \text{if } x > 0 \\ 0 & \text{otherwise,} \end{cases}$$

One checks that $\phi(x) = g(x)g(1-x)$ is a test function with the criterion of its being infinitely differentiable and vanishing outside a finite interval holding.

We know that an integrable function is not always differentiable, Sometimes we may be interested in locally integrable functions. A locally integrable function $f(x)$ defined over a certain domain (a,b) is the one for which, around every point of (a,b), there exists a neighbourhood on which it is integrable. Mathematically it means that there exist sub-intervals such as (α, β) of (a,b) such that

$$\int_\alpha^\beta |f(x)dx < \infty \tag{A.70}$$

Some examples of a locally integrable functions are continuous functions, integrable functions, a piece-wise continuous function or a constant. However, the delta function is not locally integrable. This can be understood if we consider a sequence of nested intervals $I_1, I_2, ..., I_n, ...$ i.e. each interval is included in the preceding one whose length shrinks to zero:

$$\lim_{k \to \infty} \int_{I_k} f(x)dx = 0 \tag{A.71}$$

If $\delta(x)$ replaces $f(x)$ we would get a finite value.

Formally, a generalized function or a distribution is any linear functional which is continuous on D. The set of continuous linear functionals is denoted by D^1 and its elements are called generalized functions or distributions.

A locally integrable function $f(x)$ generates a regular distribution through the following linear functional

$$(f, \phi) = \int_{-\infty}^{+\infty} f(x)\phi(x)dx \tag{A.72}$$

for all test functions $\in D$.

The linearity follows straightfowardly. For all test functions ϕ_1 and $\phi_2 \in D$ and arbitrary constants α_1 and α_2 we have

$$(f, \alpha\phi_1 + \beta\phi_2) = (\alpha \int_{-\infty}^{+\infty} f(x)\phi_1(x) + \beta \int_{-\infty}^{+\infty} f(x)\phi_2(x))dx) \tag{A.73}$$

and the right side is basically $\alpha(f, \phi_1) + \beta(f, \phi_2)$. The continuity follows from the bound

$$|f(x)| \leq \max_{(x \in supp\phi)} |\phi(x)| \int_{-\infty}^{+\infty} |f(x)| dx < \infty \qquad (A.74)$$

All ordinary continuous functions are regular generalized functions.

The delta function $\delta(x)$, however, is a singular distribution. Its support is provided by just the point $x = 0 : (supp)\delta(x) = (0)$. But we often use the symbolic operation to express

$$(\delta(x), \phi) = (\delta_{\xi=0}, \phi) = \phi(0) \qquad (A.75)$$

which in the integral form reads

$$\int \delta(x)\phi(x) dx = \phi(0) \qquad (A.76)$$

For a shifted argument the symbolic notation means

$$\int \delta(x - \xi)\phi(x) dx = \phi(\xi) \qquad (A.77)$$

We have already addressed the issue of delta-convergent sequence or simply delta sequence. In the present symbolic context we can think of the delta function as being obtained in the limit of a sequence of ordinary integrals generated by locally integrable functions $f_k(x)$

$$\int f_k(x)\phi(x) dx = \phi(0) \qquad (A.78)$$

The Gausssian, Cauchy or Lorentzian are some examples in this regard.

Finally, we can define the distributional derivative in the following way. If f and f' are two locally integrable functions then (f', ϕ) stands for

$$(f', \phi) = \int_a^b f'\phi dx = -\int_a^b f\phi' dx = -(f, \phi') \qquad (A.79)$$

where we have integrated by parts and exploited the feature of the test function so that it vanishes outside a closed, bounded interval. Thus we see that the derivative has been transferred to the test function implying the property

$$(f', \phi) = -(f, \phi'), \quad \phi \in D \qquad (A.80)$$

On integrating by parts n times, the n-th distributional derivative $f^{(n)}$ is given by

$$(f^{(n)}, \phi) = (-1)^n (f, \phi^{(n)}), \quad \phi \in D \qquad (A.81)$$

For the delta function we have symbolically the following operations for its derivative

$$\int \delta'(x - \xi)\phi(x)dx = -\phi'(\xi), \quad \phi \in D \tag{A.82}$$

and for its n-th derivative

$$\int \delta^{(n)}(x - \xi)\phi(x)dx = (-1)^n \phi^{(n)}(\xi), \quad \phi \in D \tag{A.83}$$

To illustrate how a distributional derivative might work in actual practice we take the example of the Heaviside function $H(x)$. It is locally integrable and defined by

$$H(x) = \begin{cases} +1 & \text{if } x > 0 \\ 0 & \text{if } x < 0 \end{cases}$$

Heaviside function is also called[3] the unit step function. The jumped discontinuity of $H(x)$ is obvious. It is also clear that $H(-x) = 1 = H(x)$. If the jumped discontinuity is located at the point $x = \xi$ say, then the notation $H(x - \xi)$ is used and the relation $H(\xi - x) = 1 - H(x - \xi)$ holds.

We can represent $H(x)$ by means of a distribution

$$\int H(x)\phi(x)dx = \int_0^\infty \phi(x)dx \tag{A.84}$$

The distributional derivative of H looks like

$$-\int H(x)\phi(x)dx = -\int_0^\infty \phi'(x)dx \tag{A.85}$$

where the right side is $\phi(0) - \phi(\infty) = \phi(0)$ by the definition of a test function. But $\phi(0)$ can be expressed symbolically

$$\int \delta(x)\phi(x)dx \tag{A.86}$$

So in a distributional sense the derivative[4] of $H(x)$ is given by

$$H'(x) = \delta(x) \tag{A.87}$$

Consider the following example in which a function $f(x)$ is defined by

$$f(x) = \begin{cases} x^2 & \text{if } x < a \\ x^3 & \text{if } x \ge a \end{cases}$$

where a is a constant. Expressing $f(x)$ in the manner $f(x) = x^2 + (x^3 - x^2)H(x - a)$, it is straightforward to work out the derivative $f'(x)$ in the distributional sense

---

[3] Often the notation $\theta(x)$ is used.

[4] Note that the Heaviside function is discontinuous at $x = 0$ and hence not differentiable in the ordinary sense. The derivative vanishes everywhere except at the point $x = 0$ where it does not exist.

namely, $f'(x) = 2x + (3x^2 - 2x)H(x-a) + (a^3 - a^2)\delta(x-a)$. The latter can be projected as

$$f'(x) = a^2(a-1)\delta(x-a) + \begin{cases} 2x & \text{if } x < a \\ 3x^2 & \text{if } x \geq a \end{cases}$$

When dealing with differential equations, say a PDE with an inhomogeneous term, a distribution can act for latter. In the book we have discussed at many places the role of the delta function inhomogeneity as given by the form

$$L\phi = \delta \tag{A.88}$$

The distributional solution resulting from it is the fundamental solution or Green's function.

Green's function

The use of Green's function method facilitates solving a non-homogeneous linear PDE. The idea of a Green's function goes as follows.

Let $\phi(\xi_1, \xi_2, ..., \xi_n)$ satisfy a non-homogeneous PDE in a certain n-dimensional region $\Re^n$

$$\triangle\phi \equiv (\frac{\partial^2}{\partial\xi_1{}^2} + \frac{\partial^2}{\partial\xi_2{}^2} + ... + \frac{\partial^2}{\partial\xi_n{}^2})\phi(\xi_1, \xi_2, ..., \xi_n) = \rho(\xi_1, \xi_2, ..., \xi_n) \tag{A.89}$$

where ρ is a source term. We assume that (A.89) is accompanied by the Dirichlet boundary condition on the corresponding hypersurface. We refer to $\xi \equiv (\xi_1, \xi_2, ..., \xi_n)$ as the coordinates of a variable point Q.

Consider a solution $G(\xi, \mathbf{x})$ of (A.89) when ρ is replaced by the Dirac delta function

$$\triangle G(\xi, \mathbf{x}) = \delta(\xi - \mathbf{x}) \tag{A.90}$$

where $\mathbf{x}$ corresponds to a fixed point P in Ω. Like ϕ, G also obeys the Dirchlet boundary condition on the hypersurface. $G(\xi, \mathbf{x})$ is called the Green's function of the operator $\triangle$.

It is easy to check that ϕ as given by the following integral

$$\phi(\xi) = \int_\Omega G(\xi, \mathbf{x})\rho(\mathbf{x})d\mathbf{x} \tag{A.91}$$

satisfies (A.90), by operating on both sides of (A.91), by $\triangle$ and using (A.89).

We state the Dirichlet boundary problem for ϕ as follows

$$\triangle\phi(\xi) = \rho(\xi), \quad \xi \in \Omega, \quad \phi(\xi) = 0, \quad \xi \in \partial\Omega \tag{A.92}$$

Note that the vanishing of ϕ on the boundary is consistent with the vanishing of G on the same from the given integral representation (A.91) of ϕ in terms of G.

Appendix B

Fourier transform

Fourier transform

A periodic function with period $2l$ can be represented by a Fourier series in the complex form [1]

$$f(x) = \sum_{n=-\infty}^{\infty} a_n e^{\frac{in\pi x}{l}} \tag{B.1}$$

where the complex coefficients are given by the integral

$$a_n = \frac{1}{2l} \int_{-l}^{+l} f(z) e^{-\frac{in\pi z}{l}} dz \tag{B.2}$$

z being a dummy variable.

Letting $(-l, +l) - > (-\infty, +\infty)$ we can think of connecting to functions which are not periodic. Towards this end, we define

$$p_n = \frac{n\pi}{l} \quad \Rightarrow \quad \triangle p \equiv p_{n+1} - p_n = \frac{\pi}{l} \quad \Rightarrow \frac{1}{2l} = \frac{\triangle p}{2\pi} \tag{B.3}$$

which puts (B.1) and (B.2) in the forms

$$f(x) = \sum_{n=-\infty}^{\infty} a_n e^{ip_n x} \tag{B.4}$$

along with

$$a_n = \frac{\triangle p}{2\pi} \int_{-l}^{+l} f(z) e^{-ip_n z} dz \tag{B.5}$$

---

[1] It is well known that a periodic function of period 2π can be represented by the expansion $f(x) = \frac{a_0}{2} + \sum_{n=-\infty}^{\infty} a_n \cos(nx) + b_n \sin(nx)$. Noting that $(\cos(t), \sin(t)) = \frac{e^{it} \pm e^{-it}}{2}$, it straightforwardly follows, on setting $d_0 = \frac{a_0}{2}, d_n = \frac{a_n - ib_n}{2}, d_{-n} = \frac{a_n + ib_n}{2}$, that $f(x) = \sum_{n=-\infty}^{\infty} d_n e^{inx}$.

As a consequence we can express

$$f(t) = \sum_{n=-\infty}^{\infty} [\frac{\triangle p}{2\pi} \int_{-l}^{+l} f(z)e^{-ip_n z}dz]e^{ip_n t} \tag{B.6}$$

where we have used the symbol t in place of x to avoid confusion. $f(t)$ can be put in the form of the sum

$$f(t) = \frac{1}{2\pi} \sum_{n=-\infty}^{\infty} G(p_n)\triangle p \tag{B.7}$$

where

$$G(p_n) = \int_{-l}^{+l} f(z)e^{ip_n(t-z)}dz \tag{B.8}$$

Of course we can transform the infinite sum to an integral by letting $l \to \infty$ which implies by (B.3) $\triangle p \to 0$. This gives us from (B.8)

$$G(p) = \int_{-\infty}^{\infty} f(z)e^{ip(t-z)}dz \tag{B.9}$$

where p now represents a continuous variable and the suffix n on p put previously (which indicted discreteness) has been dropped.

The summation in (B.7) is also changed to an indefinite integral and $f(t)$ reads

$$f(t) = \frac{1}{2\pi} \int_{-\infty}^{\infty} G(p)dp = \frac{1}{2\pi} \int_{-\infty}^{\infty} e^{ipt} dp \int_{-\infty}^{\infty} f(z)e^{-ipz}dz \tag{B.10}$$

Denoting $\hat{f}(p)$ to stand for the integral

$$\hat{f}(p) = \int_{-\infty}^{\infty} f(z)e^{-ipz}dz \tag{B.11}$$

$f(t)$ takes the form

$$f(t) = \frac{1}{2\pi} \int_{-\infty}^{\infty} \hat{f}(p)e^{ipt}dp \tag{B.12}$$

We call $\hat{f}(p)$ to be the Fourier transform of $f(t)$ and (B.12) as the inversion formula or inversion theorem for Fourier transform. Note that the Fourier transformation is a linear transformation.

From (B.11) and (B.12) it is easy to convince oneself that the Fourier transform of 1 is $2\pi\delta(p)$ while the Fourier transform of the Heaviside step function is $\frac{1}{ip}+\pi\delta(p)$.

For a complex function f, by separating into its real and imaginary parts, it can be easily shown that the Fourier transforms of the complex conjugate (denoted by overlines) of $f(t)$ and $f(-t)$ are respectively

$$F[\overline{f(t)}] = \overline{\hat{f}(-p)}, \quad F[\overline{f(-t)}] = \overline{\hat{f}(p)} \quad \text{for all } p \tag{B.13}$$

For the existence of a Fourier transform certain conditions need to be satisfied which we state below. These are called Dirichlet's conditions. First, $f(t)$ has to be absolutely integrable. In other words, $\int_{-\infty}^{\infty} |f(t)|dt$ converges. Second, f has a finite number of points of maxima and minima in any finite interval. This means that functions like $f(t) = \sin(\frac{1}{t})$ are excluded from consideration. Third, $f(t)$ has only a finite number of points of discontinuities in any finite interval. This means that functions such as

$$f(t) = \begin{cases} 0 & \text{if } t \text{ rational} \\ 1 & \text{if } t \text{ irrational} \end{cases}$$

are disregarded from consideration. In this way, Dirichlet's conditions give a set of sufficient conditions for the existence of a Fourier transform.

In some cases the factor $\frac{1}{2\pi}$ is symmetrized between the Fourier transform $\hat{f}(p)$ and its inverse to read as a pair

$$\hat{f}(p) \quad = \quad \frac{1}{\sqrt{2\pi}} \int_{-\infty}^{\infty} f(z)e^{-ipz}dz \tag{B.14}$$

$$f(t) \quad = \quad \frac{1}{\sqrt{2\pi}} \int_{-\infty}^{\infty} \hat{f}(p)e^{ipt}dp \tag{B.15}$$

In this book we have adopted such a convention.

Let us consider the following examples.

(1) Let $f(t)$ be defined by

$$f(t) = e^{-\gamma|t|}, \quad \gamma > 0$$

Then its Fourier transform is given by

$$\hat{f}(p) = \frac{1}{\sqrt{2\pi}} \int_{-\infty}^{\infty} e^{-\gamma|t|}e^{-ipt}dt$$

The above can be expressed as

$$\hat{f}(p) = \frac{1}{\sqrt{2\pi}}2 \lim_{R \to \infty} \int_{0}^{R} e^{-\gamma t}\cos(pt)dt$$

where because of its odd character $\sin(pt)$ term vanishes. Performing the integral and taking the limit $R \to \infty$ we find

$$\hat{f}(p) = \sqrt{\frac{2}{\pi}}\frac{\gamma}{p^2 + \gamma^2}$$

(2) Let $f(t)$ be given by the Heaviside function

$$f(t) = H(a - |t|), \quad a > 0$$

i.e.

$$f(t) = \begin{cases} 0 & \text{if } |t| > a \\ 1 & \text{if } |t| < a \end{cases}$$

Then the Fourier transform of $f(t)$ is given by

$$\hat{f}(p) = \frac{1}{\sqrt{2\pi}} \int_{-\infty}^{\infty} H(a - |t|) e^{-ipt} dt$$

which reduces to simply

$$\hat{f}(p) = \frac{1}{\sqrt{2\pi}} \int_{-a}^{a} e^{-ipt} dt$$

Evaluating the integral it turns out that

$$\hat{f}(p) = \sqrt{\frac{2}{\pi}} \frac{\sin(pa)}{p}$$

Depending on the odd-ness and even-ness of a function it proves useful to define the sine transform and cosine transform of the function respectively. The Fourier sine transform and its inverse are defined by

$$f_s(p) \;\; = \;\; \sqrt{\frac{2}{\pi}} \int_0^{\infty} f(z) \sin(pz) dz, \quad 0 \le p < \infty \tag{B.16}$$

$$f_s(t) \;\; = \;\; \sqrt{\frac{2}{\pi}} \int_0^{\infty} f_s(p) \sin(pt) dp, \quad 0 \le t < \infty \tag{B.17}$$

On the other hand, the Fourier cosine transform and its inverse are defined by

$$f_c(p) \;\; = \;\; \sqrt{\frac{2}{\pi}} \int_0^{\infty} f(z) \cos(pz) dz \quad 0 \le p < \infty \tag{B.18}$$

$$f_c(t) \;\; = \;\; \sqrt{\frac{2}{\pi}} \int_0^{\infty} f_s(p) \cos(pt) dp, \quad 0 \le t < \infty \tag{B.19}$$

where we have assumed Fourier sine and cosine transforms to exist which is ensured by assuming $\int_0^{\infty} f(t) dt$ to be absolutely convergent.

The linearity of Fourier sine and cosine transforms is self-evident. In the following we work out the formula for the derivatives. Let us assume that a function $f(t)$ is differentiable n times with respect to t such that $f(t)$ and its $(n-1)$-derivatives with respect to t all tend to zero as $t \to \infty$. The Fourier sine transform for the first derivative is given by

$$[f_x]_s(p) = \sqrt{\frac{2}{\pi}} \int_0^{\infty} f'(z) \sin(pz) dz \tag{B.20}$$

which when integrated by parts yields

$$[f_x]_s(p) = -p\sqrt{\frac{2}{\pi}} \int_0^\infty f(z)\cos(pz)dz = -pf_c(p) \tag{B.21}$$

by the assumed conditions on $f(t)$.
Integrating again we find

$$[f_{xx}]_s(p) = \sqrt{\frac{2}{\pi}}pf(0) - p^2 f_s(p) \tag{B.22}$$

Similarly for the first derivative of the cosine transform the following holds

$$[f_x]_c(p) = pf_s(p) - \sqrt{\frac{2}{\pi}}f(0) \tag{B.23}$$

and for the second derivative we have the result

$$[f_{xx}]_c(p) = -p^2 f_c(p) - \sqrt{\frac{2}{\pi}}f'(0) \tag{B.24}$$

More generally, we can derive for multiple derivatives the formulae

$$(f_{xx\ldots(2n)-\text{times}})_c(p) = -\sum_{m=0}^{n-1}(-1)^m a_{2n-2m-1}p^{2m} + (-1)^n p^{2n} f_c(p) \tag{B.25}$$

$$(f_{xx\ldots(2n+1)-\text{times}})_c(p) = -\sum_{m=0}^{n}(-1)^m a_{2n-2m}p^{2m} + (-1)^n p^{2n+1} f_s(p) \tag{B.26}$$

$$(f_{xx\ldots(2n)-\text{times}})_s(p) = -\sum_{m=0}^{n}(-1)^m a_{2n-2m}p^{2m-1} + (-1)^n p^{2n} f_s(p) \tag{B.27}$$

$$(f_{xx\ldots(2n+1)-\text{times}})_s(p) = -\sum_{m=1}^{n}(-1)^m a_{2n-2m+1}p^{2m-1} + (-1)^n p^{2n+1} f_c(p) \tag{B.28}$$

where the quantity a_r denotes

$$a_r = \sqrt{\frac{2}{\pi}} \lim_{t \to 0^+} f^r(t) \tag{B.29}$$

As simple applications of sine and cosine transforms we note that since

$$f_s[f(\alpha t)] = \sqrt{\frac{2}{\pi}} \int_0^\infty f(\alpha t)\sin(pt)dt \tag{B.30}$$

$$f_c[f(\alpha t)] = \sqrt{\frac{2}{\pi}} \int_0^\infty f(\alpha t)\cos(pt)dt \tag{B.31}$$

we have directly

$$f_s(e^{-bt}) = \sqrt{\frac{2}{\pi}} \int_0^\infty e^{-bt} \sin(pt)dt = \sqrt{\frac{2}{\pi}} \frac{b}{b^2 + p^2} \tag{B.32}$$

$$f_c(e^{-bt}) = \sqrt{\frac{2}{\pi}} \int_0^\infty e^{-bt} \cos(pt)dt = \sqrt{\frac{2}{\pi}} \frac{p}{b^2 + p^2} \tag{B.33}$$

Similarly when a decaying exponential is appended

$$f_s(e^{-bt}\cos(at)) = \frac{1}{2}\sqrt{\frac{2}{\pi}}[\frac{p-a}{b^2 + (p-a)^2} + \frac{p+a}{p^2 + (p+a)^2}] \tag{B.34}$$

$$f_c(e^{-bt}\cos(at)) = \frac{b}{2}[\frac{1}{b^2 + (a-p)^2} + \frac{1}{b^2 + (a+p)^2}] \tag{B.35}$$

In both cases above $a, b > 0$ and $t \geq 0$.

As another example consider the exponential function $f(x) = e^{-\frac{x}{a}}$, $x > 0$ where $a > 0$. That the Fourier transform is a complex function can be seen from a direct evaluation

$$\hat{f}(p) = \int_0^\infty e^{-\frac{x}{a}} e^{-ipx} dx = \frac{a}{1 + ipa} \tag{B.36}$$

This implies that

$$|\hat{f}(p)|^2 = \frac{a^2}{1 + p^2 a^2} \tag{B.37}$$

which corresponds to a bell-shaped Lorentz curve discussed in Appendix A. We furnish in Table B.1 certain standard results of Fourier transform.

TABLE B.1: Some standard results in Fourier transform.

Function	Fourier transform
$\sqrt{2\pi}f(t) = \int_{-\infty}^{\infty} \hat{f}(p)e^{ipt}dp$	$\sqrt{2\pi}\hat{f}(p) = \int_{-\infty}^{\infty} f(z)e^{-ipz}dz$
1	$2\pi\delta(p)$
$f'(t)$	$ip\hat{f}(p)$
$f(t - t_0)$	$\hat{f}(p)e^{-ipt_0}$
$\cos(\lambda t)$	$\pi[\delta(p - \lambda) + \delta(p + \lambda)]$
$\sin(\lambda t)$	$-i\pi[\delta(p - \lambda) - \delta(p + \lambda)]$
$e^{-|\gamma||t|}$	$\sqrt{\frac{2}{\pi}}\frac{\gamma}{p^2+\gamma^2}$
$\frac{1}{t^2+a^2}$	$\frac{1}{a}\sqrt{\frac{\pi}{2}}e^{-a|p|}, \quad a > 0$
$e^{-at^2}, \quad a > 0$	$\frac{1}{\sqrt{2a}}e^{-\frac{p^2}{4a}}$
$\delta(x)$	$\frac{1}{\sqrt{2\pi}}$
$H(a - |t|), \quad a > 0$	$\sqrt{\frac{2}{\pi}}\frac{\sin px}{p}$
$\sum_{n=-\infty}^{\infty}\delta(t - nT)$	$\frac{\sqrt{2\pi}}{T}\sum_{m=-\infty}^{\infty}\delta(p - \frac{2m\pi}{T})$

Convolution theorem and Parseval relation

To derive the convolution theorem for the Fourier transform let us define the convolution of two functions $f(t)$ and $g(t)$ as the integral

$$\int_{-\infty}^{+\infty} f(\tau)g(t-\tau)d\tau = f * g \tag{B.38}$$

where $f * g \ (= g * f)$ speaks of the convolution integral of Fourier transform.

Let the functions f and g have their Fourier transforms as $\tilde{f}(p)$ and $\tilde{g}(p)$ respectively. Then

$$f * g = \frac{1}{\sqrt{2\pi}} \int_{-\infty}^{\infty} f(\tau)d\tau \int_{-\infty}^{\infty} \tilde{g}(p)e^{ip(t-\tau)}dp \tag{B.39}$$

where $\tilde{g}(p)$ is the Fourier transform of $g(t-\tau)$. Writing

$$f * g = \int_{-\infty}^{\infty} \tilde{g}(p)e^{ipt}dp\left(\frac{1}{\sqrt{2\pi}}\int_{-\infty}^{\infty} f(\tau)e^{-ip\tau}d\tau\right) \tag{B.40}$$

it transpires that

$$f * g = \int_{-\infty}^{\infty} \tilde{f}(p)\tilde{g}(p)e^{ipt}dp \tag{B.41}$$

where $\tilde{f}(p)$ is the Fourier transform of $f(t)$. This is one form of convolution theorem showing that $f * g$ is the inverse Fourier transform of $\sqrt{2\pi}\tilde{f}(p)\tilde{g}(p)$.

Let us now put $t = 0$. From (B.40) and (B.42) we arrive at the form

$$\int_{-\infty}^{+\infty} f(\tau)g(-\tau)d\tau = \int_{-\infty}^{\infty} \tilde{f}(p)\tilde{g}(p)dp \tag{B.42}$$

If we put in place of $g(-\tau)$ the function $\overline{g(\tau)}$ and note that $F[\overline{g(-t)}] = \overline{\tilde{g}(p)}$ we deduce

$$\int_{-\infty}^{+\infty} f(\tau)\overline{g(\tau)}d\tau = \int_{-\infty}^{\infty} \tilde{f}(p)\overline{\tilde{g}(p)}dp \tag{B.43}$$

If f and g are equal i.e. $f(t) = g(t)$ and $\tilde{f}(p)$ is the common Fourier transform then

$$\int_{-\infty}^{+\infty} f(t)\overline{f(t)}dt = \int_{-\infty}^{\infty} \tilde{f}(p)\overline{\tilde{f}(p)}dp \tag{B.44}$$

which implies

$$\int_{-\infty}^{+\infty} |f(t)|^2 dt = \int_{-\infty}^{\infty} |\tilde{f}(p)|^2 dp \tag{B.45}$$

(B.43) - (B.45) are called Parseval's relation.

As an application of one of Parseval's relations let us consider the function $f(x)$ as defined by

$$f(x) = \begin{cases} 1 - |x| & \text{in (-1,1)} \\ 0 & \text{in } |x| > 1 \end{cases}$$

Here the Fourier transform of $f(x)$ is given by

$$\tilde{f}(p) = \frac{1}{\sqrt{2\pi}} \int_{-1}^{0} (1+x)e^{-ipx}\,dx + \frac{1}{\sqrt{2\pi}} \int_{0}^{1} (1-x)e^{-ipx}\,dx \qquad \text{(B.46)}$$

Evaluation of the integrals gives

$$\tilde{f}(p) = \frac{1}{\sqrt{2\pi}} [\frac{\sin(\frac{p}{2})}{\frac{p}{2}}]^2 \qquad \text{(B.47)}$$

Applying now the Parseval's relation (B.47)

$$\int_{-\infty}^{+\infty} [\frac{\sin(\frac{p}{2})}{\frac{p}{2}}]^4\,dp = 2\pi \int_{-1}^{1} (1-|x|)^2\,dx = 4\pi \int_{0}^{1} (1-x)^2\,dx = \frac{4\pi}{3} \qquad \text{(B.48)}$$

Appendix C

Laplace transform

Laplace transform

For any function $f(t)$, which is

(i) piecewise continuous in the closed interval $[0, a], a > 0$ and
(ii) obeys $|f(t)| \leq Ke^{\alpha t}$, where $t > T$, K, α and T are constants with $K, T > 0$,
i.e. $f(t)$ is of exponential order as $t \to \infty$
then the Laplace transform of $f(t)$ is said to exist which is a function of a complex
variable s and is given by

$$F(s) = \frac{1}{2i\pi} \int_0^\infty e^{-st} f(t) dt, \quad Re(s) > \alpha \tag{C.1}$$

The above proposition is only a sufficient condition for the existence of the Laplace
transform but by no means necessary. This basically means that a function may
have a Laplace transform without satisfying the above two conditions. Logarithmic
function and unit impulse function are examples in this regard. For the logarithmic function the result is $L(\ln(t)) = \Gamma'(1) - \ln(s)$, where Γ is the Eulerian gamma
function while for the unit impulse function, which is essentially the delta function,
the Laplace transform is unity. The integral (C.1) does not converge for functions
that increase faster than the exponential and these functions do not have a Laplace
transform.

Laplace transform of the derivatives is easily worked out in the following forms.
For the first and second derivatives the expressions are

$$L[f'(t)] = sL[f(t)] - f(0) \tag{C.2}$$

and

$$L[f''(t)] = s^2 L[f(t)] - sf(0) - f'(0) \tag{C.3}$$

while for the n-th derivative it is given by

$$L[f^{(n)}(t)] = s^n L[f(t)] - s^{n-1}f(0) - s^{n-2}f'(0) - f^{n-1}(0) \qquad \text{(C.4)}$$

The convolution theorem for the Laplace transform states that if $L[f(t)] = F(s)$ and $L[g(t)] = G(s)$ then

$$L[\int_0^t f(t)g(t-\tau)d\tau] = F(s)G(s) \qquad \text{(C.5)}$$

where the integral stands for the convolution integral of Laplace transform.

We furnish in Table C.1 certain standard results of Laplace transform.

Inversion theorem for Laplace transform

Here the main aim is to construct $f(t)$ from the form of $F(s)$. The inversion theorem for Laplace transform which is also referred to as the Mellin's inverse formula or Bromwich integral gives sufficient conditions under which $F(s)$ is the Laplace transform of $f(t)$.

Let $F(s)$ be an analytic function of a complex variable s and of order $O(s^{-k}), k > 1$ for $r = Re(s) > \alpha$. Further, $F(s)$ is real for $r = Re(s) > \alpha$. Then an integral formula for the inverse Laplace transform is given by

$$f(t) = \frac{1}{2i\pi} \int_{r-i\infty}^{r+i\infty} e^{st} F(s)ds, \quad Re(s) > \alpha \qquad \text{(C.6)}$$

As would be realized in solving problems of inverse formula Cauchy's residue theorem greatly facilitates evaluation of the complex integral. It can be shown that the function $f(t)$ is independent of r at points of continuity of $f(t)$ whenever $r > a$ and has the following properties

(i) $f(t)$ is continuous for all t,

(ii) f(t) = 0 for $t < 0$,

(iii) $f(t)$ is of exponential order i.e. $O(e^{\alpha t})$ at $t \to \infty$ and

(iv) $F(s)$ is the Laplace transform of $f(t)$.

Asymptotic form for Laplace's inversion integral

To evaluate the asymptotic form (i.e. for large t) for the inversion integral

$$f(t) = \frac{1}{2i\pi} \int_{r-i\infty}^{r+i\infty} e^{st} F(s) ds, \quad 0 < \beta < 1 \tag{C.7}$$

from the behaviour of $F(s)$ near its singularity with the largest real part, let $s = x+iy$ and the singularity of $F(s)$ be located at the point $s = s_0 = a + ib$ where $a < r$. Subject to the following assumptions that

(i) $F(s)$ is analytic in the region where $Re(s) \geq a - \delta$, $\delta > 0$ except at $s = s_0$,
(ii) $F(s) \to 0$ uniformly with x for $a - \delta \leq x < r$ as $y \to \pm\infty$,
(iii) $\int^y |F(s)| dy$ converges for $y \to \pm\infty$ for $a - \delta \leq x < r$ and
(iv) near $s = s_0$, $F(s)$ can be expanded as

$$F(s) = \sum_{n=-1}^{\infty} a_n (s - s_0)^n + (s - s_0)^{-\beta} \sum_{n=0}^{\infty} b_n (s - s_0)^n, \quad 0 < \beta < 1 \tag{C.8}$$

where the two series converge for $0 < |s - s_0| \leq l$, $f(t)$ has, for large t, the following representation

$$f(t) \sim e^{s_0 t} [a_{-1} + \frac{\sin\beta\pi}{\pi} \sum_{n=0}^{\infty} (1)^n b_n \Gamma(n + 1 - \beta) \frac{1}{t^{n+1-\beta}}] \tag{C.9}$$

Before we outline the proof, it may be mentioned that the above result also addresses the particular case of Watson lemma which states that if $F(s)$ can be expanded as

$$F(s) \sim \sum_{n=0}^{\infty} b_n \frac{1}{s^{\beta-n}}, \quad Re(\beta) > 1 \tag{C.10}$$

as $s- > 0^+$, then the asymptotic expansion of $f(t)$ around $t = 0$ is provided by

$$f(t) \sim \sum_{n=0}^{\infty} b_n \Gamma(n + 1 - \beta) \frac{1}{t^{n+1-\beta}} \quad Re(\beta) > 1 \tag{C.11}$$

Watson's lemma has applications in problems dealing with the asymptotic nature of integrals.

To get along with the proof of the expansion (C.8), we focus on the contour C as sketched in Figure C.1.

C is defined by the line segments LM, QR which are portions of the straight line $x = a - \delta$ (note $s = x + iy$), parallel segments MN, PQ and a small circle c_ρ, ρ being its radius fulfilling $s - s_0 = \rho \equiv \frac{\epsilon}{t}$, ϵ being an infinitesimal quantity.

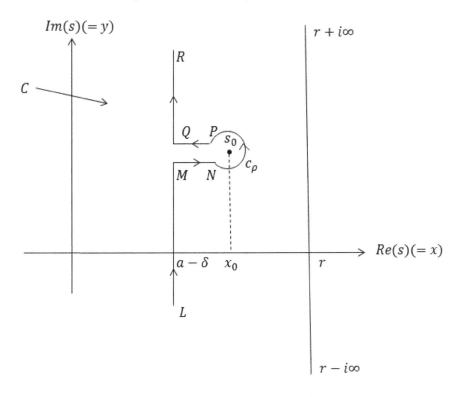

FIGURE C.1: The contour C.

For the validity of the expansion (C.7) over MN, PQ and c_ρ, we choose $\delta < l$. Because of the assumptions made in (i) and (ii), we replace $(r - i\infty, r + i\infty)$ by the contour C to write

$$e^{-s_0 t} f(t) = \frac{1}{2i\pi} \int_C e^{(s-s_0)t} F(s) ds, \quad 0 < \beta < 1$$

Taking into account the various components of the contour C as shown in Figure C.1, we split the right side of (C.11) into the following parts I_1, I_2 and I_3

$$I_1 = \frac{1}{2i\pi} \int_{LM+QR} e^{(s-s_0)t} F(s) ds, \quad 0 < \beta < 1 \tag{C.12}$$

$$I_2 = \frac{1}{2i\pi} \int_{MN+PQ} e^{(s-s_0)t} F(s) ds, \quad 0 < \beta < 1 \tag{C.13}$$

$$I_3 = \frac{1}{2i\pi} \int_{C_\rho} e^{(s-s_0)t} F(s) ds, \quad 0 < \beta < 1 \qquad \text{(C.14)}$$

For I_1 we see that

$$|I_1| \le \frac{1}{2\pi} \int_{-\infty}^{b-\frac{\epsilon}{t}} e^{-\delta t} |F(s)| ds + \frac{1}{2\pi} \int_{b+\frac{\epsilon}{t}}^{\infty} e^{-\delta t} |F(s)| ds$$

which because of the condition (iii) reduces to

$$|I_1| \le K e^{-\delta t}$$

where K is a constant.

For I_3 we see that

$$I_3 = \frac{1}{2\pi} \int_0^{2\pi} e^{\rho t e^{i\theta}} \left[\sum_{n=-1}^{\infty} a_n \rho^n e^{in\theta} + \rho^{-\beta} e^{-i\beta\theta} \sum_{n=0}^{\infty} b_n \rho^n e^{in\theta} \right] \rho e^{i\theta} d\theta$$

which as ρ tends to zero gives

$$\lim_{\rho \to 0} I_3 = \frac{1}{2\pi} \int_0^{2\pi} a_{-1} d\theta = a_{-1}$$

Finally, for I_2, since on MN, $s = s_0 + xe^{-i\pi}$ and on PQ, $s = s_0 + xe^{i\pi}$, we deduce

$$I_2^{MN} = \frac{1}{2i\pi} \int_{\frac{\epsilon}{t}}^{\delta} e^{-xt} \left[\sum_{n=-1}^{\infty} a_n (-1)^n x^n + x^{-\beta} e^{i\pi\beta} \sum_{n=0}^{\infty} b_n (-1)^n x^n \right] dx$$

and

$$I_2^{PQ} = -\frac{1}{2i\pi} \int_{\frac{\epsilon}{t}}^{\delta} e^{-xt} \left[\sum_{n=-1}^{\infty} a_n (-1)^n x^n + x^{-\beta} e^{-i\pi\beta} \sum_{n=0}^{\infty} b_n (-1)^n x^n \right] dx$$

On putting $xt = z$ the above two can be combined to represent $I_2 = I_2^{MN} + I_2^{PQ}$ as

$$I_2 = \frac{\sin \pi\beta}{\pi} \int_{\epsilon}^{\delta t} dz \frac{e^{-z}}{z} \sum_{n=0}^{\infty} b_n (-1)^n \left(\frac{z}{t}\right)^{n+1-\beta}$$

Collecting the above evaluations of I_1, I_2 and I_3 we find on proceeding to the limit $\epsilon \to 0$

$$e^{-s_0 t} f(t) \sim a_{-1} + \frac{\sin \pi\beta}{\pi} \int_0^{\delta t} dz \frac{e^{-z}}{z} \sum_{n=0}^{\infty} b_n (-1)^n \left(\frac{z}{t}\right)^{n+1-\beta}$$

We now proceed to show that as $t \to \infty$ the quantity ζ defined by

$$\zeta = [\int_0^{\delta t} dz \frac{e^{-z}}{z} \sum_{n=0}^{\infty} b_n (-1)^n (\frac{z}{t})^{n+1-\beta} - \sum_{n=0}^{\infty} (-1)^n b_n \frac{\Gamma(n+1-\beta)}{t^{n+1-\beta}}] t^{n+1-\beta} \quad \text{(C.15)}$$

goes to 0. Towards this end, we use the integral representation of the gamma function

$$\Gamma(\lambda+1) = \int_0^{\infty} e^{-t} t^{\lambda} dt, \quad \lambda > -1$$

to re-write ζ as

$$\zeta = t^{n+1-\beta} [\int_0^{\delta t} dz \frac{e^{-z}}{z} \sum_{m=0}^{\infty} b_m (-1)^m (\frac{z}{t})^{m+1-\beta} - \sum_{m=0}^{n} b_m (-1)^m \int_0^{\infty} dz \frac{e^{-z}}{z} (\frac{z}{t})^{m+1-\beta}]$$

Splitting the integral appearing in the second term into portions $(0, \delta t)$ and $(\delta t, \infty)$ and subtracting the piece of summation over $m=0$ to n from the summation in the first integral we can write

$$\zeta = t^{n+1-\beta} [\int_0^{\delta t} dz \frac{e^{-z}}{z} \sum_{m=n+1}^{\infty} b_m (-1)^m (\frac{z}{t})^{m+1-\beta}$$

$$- \sum_{m=0}^{n} b_m (-1)^m \int_{\delta t}^{\infty} dz \frac{e^{-z}}{z} (\frac{z}{t})^{m+1-\beta}]$$

This can be expressed in the manner

$$\zeta = t^{-1} I - \sum_{m=0}^{n} I'_m$$

where I and I'_m stand for

$$I = \int_0^{\delta t} dz \, z^{n+1-\beta} e^{-z} \sum_{m=n+1}^{\infty} b_m (-1)^m (\frac{z}{t})^{m-n-1}$$

and

$$I'_m = t^{n-\beta} e^{-\delta t} \sum_{m=0}^{n} b_m (-1)^m \int_0^{\infty} dx (\delta + \frac{x}{t})^{m-\beta} e^{-x}$$

where we have put $z = \delta t + x$. It is clear that $I'_m \to 0$ for $m = 1, 2, ..., n$.
About I we find

$$|I| \leq \int_0^{\delta t} dz \, z^{n+1-\beta} e^{-z} \sum_{m=n+1}^{\infty} |b_m| (\frac{z}{t})^{m-n-1}$$

$$\leq \int_0^{\delta t} dz \, z^{n+1-\beta} e^{-z} \sum_{m=n+1}^{\infty} |b_m| \delta^{m-n-1}$$

$$\leq B \int_0^{\delta t} dz\, z^{n+1-\beta} e^{-z}$$

$$\leq B\Gamma(n+2-\beta)$$

In the above steps we have noted that the series $\sum_n b_n (s - s_0)^n$ is convergent and its radius of convergence is $< \delta$. Also $\int_0^{\delta t} \leq \int_0^\infty$ and B is a constant. Since there is a factor t^{-1} in (C.28) it follows that $\zeta \to 0$ as $t \to \infty$.

From (C.23), (C.24) and using the limiting result of ζ we arrive at (C.9).

TABLE C.1: Some standard results of Laplace transform. In the text we have used the notation $\tilde{f}$ in place of F. Note that $F(s) = L(f(t))$ and $f(t) = L^{-1}F(s)$.

$f(t)$	$F(s), \quad s \in \mathbb{C}$
1	$\frac{1}{s}$
$e^{\alpha t}$	$\frac{1}{s-\alpha}$
t^n	$\frac{n!}{s^{n+1}}, \quad n = 0, 1, 2, \ldots$
t^α	$\frac{\Gamma(\alpha+1)}{s^{\alpha+1}}, \quad \alpha \geq 0$
$f'(t)$	$sL(f) - f(0)$
$f''(t)$	$s^2 L(f) - sf(0) - f'(0)$
$f(t - t_0)$	$F(s)e^{-st_0}$
$\cos(\alpha t)$	$\frac{s}{s^2+\alpha^2}$
$\sin(\alpha t)$	$\frac{\alpha}{s^2+\alpha^2}$
$\cosh(\alpha t)$	$\frac{s}{s^2-\alpha^2}$
$\sinh(\alpha t)$	$\frac{\alpha}{s^2-\alpha^2}$
$\delta(t - \alpha)$	$e^{-\alpha s}$
$\frac{ae^{-\frac{a^2}{4t}}}{2\sqrt{\pi t}}$	$\frac{e^{-a\sqrt{s}}}{\sqrt{s}}$
$\frac{ae^{-\frac{a^2}{4t}}}{2t\sqrt{\pi t}}$	$e^{-a\sqrt{s}}$

Bibliography

1. Bagchi, B., (2017) *Advanced Classical Mechanics*, CRC/Taylor and Francis, London.
2. Balakrishnan, V., (2003) All about the Dirac delta function, *Resonance*, Volume 8, Issue 8, 48.
3. Bitsadze, A.V. and Kalinichenko, D.F., (1983) *A Collection of Problems on the Equations of Mathematical Physics*, MIR Publishers, Moscow.
4. Boas, M.L., (1966) *Mathematical Methods in the Physical Sciences*, John Wiley & Sons, Inc., New York.
5. Corinaldesi, E., (1998) *Classical Mechanics for Physics Graduate Students*, World Scientific, Singapore.
6. Costin, O., (2008) *Asmptotics and Borel Summability*, Chapman and Hall/CRC, London.
7. Dijk, G. van, (2013) *Distribution Theory*, Walter de Gruyter GmbH, Berlin.
8. James, J.F., (2011) *A Student's Guide to Fourier Transforms*, Cambridge University Press, Cambridge.
9. Joglekar, S.D., (2006) *Mathematical Physics: Advanced Topics*, Universities Press, India.
10. John, F., (1982) Partial Differential Equations, Applied Mathematics Series, Springer, New York.
11. Johnson, S.G., (2017) *When functions have no value(s): Delta functions and distributions*, MIT Course 18.303 Notes.
12. Kanwal, R.P., (2004) *Generalized Functions - Theory and Applications*, Birkhäuser Boston.
13. Kersale, E., (2003) *Analytic Solutions of Partial Differential Equations*, http://www.maths.leeds.ac.uk/~kersale/
14. Kinnunen, J., (2017) *Partial Differential Equations*, Lecture notes, Aalto University, Finland.
15. Kumaran, V., *Lecture Notes on Steady and Unsteady Diffusion*. IISc., Bangalore.
16. Lamoureux, M.P., (2006) *The Mathematics of Partial Differential Equations and the Wave Equation*, Lecture at Seismic Imaging Summer School (unpublished).
17. Lass, H.,(1950) *Vector and Tensor Analysis*, McGraw-Hill, New York.
18. Lindenbaum, S.D., (1996) *Mathematical Methods in Physics*, World Scientific.
19. Logan, J.D., (2006) *Beginning Partial Differential Equations*, John Wiley & Sons., Inc., New York.
20. Muthukumar, T., (2014) *Partial Differential Equations*, Indian Institute of Technology, Kanpur, Lecture notes.

21. Olver, P., (2013) *Introduction to Partial Differential Equations*, Springer, Berlin.

22. O'Neil, P.V., (1999)*Begining Partial Differential Equations*, John Wiley & Sons, Inc., New York.

23. Petrovsky, I.G., (1991) *Lectures on Partial Differential Equations*, Dover Publications, Inc., New York.

24. Sneddon, I.N., (1957) *Elements of Partial Differential Equations*, McGraw-Hill, New York.

25. Sommerfeld, A., (1949) *Partial Differential Equations in Physics*, Academic Press, New York.

26. Stavroulakis, I.P. and Tersian, S.A., (2004) *Partial Differential Equations: An Introduction with Mathematica and MAPLE*, World Scientific, Singapore.

27. Strauss, W.A., (1992) *Partial Differential Equations, An Introduction*, John Wiley & Sons., Inc., New York.

28. Williams, W.E., (1980) *Partial Differential Equations*, Oxford University Press, Oxford.

29. Yanovsky, I., (2005) *Partial Differential Equations: Graduate Level Problems and Solutions* (unpublished).

30. Zachmann, D. and Duchateau P., *Schaum's Outline of Partial Differential Equations*, McGraw-Hill, New York.

Index

For Product Safety Concerns and Information please contact our EU
representative GPSR@taylorandfrancis.com Taylor & Francis Verlag GmbH,
Kaufingerstraße 24, 80331 München, Germany

Printed and bound by CPI Group (UK) Ltd, Croydon, CR0 4YY
01/05/2025
01858518-0007